ADVANCE PRAISE FOR
THE AGE OF AGILE

"At a time when the quest for shareholder value is leading many companies to shortchange their customers, employees, and everyone else, Steve Denning reminds us that it doesn't have to be this way. Rich with examples of Agile management at work, *The Age of Agile* envisions an economy that is more innovative, more humane, and more inclusive."

—LYNN S. PAINE,
John G. McLean Professor of Business Administration,
Harvard Business School, and coauthor of *Capitalism at Risk:*
Rethinking the Role of Business

"An invaluable guide to transforming the old—unadaptive hierarchical management—into the new—the customer-involved, team-empowered company capable of continuously re-inventing itself."

—ROBERT M. RANDALL,
Editor-in-Chief, *Strategy & Leadership*, and editor/author of
The Portable MBA in Strategy

"Steve Denning's *The Age of Agile* is a call to arms for why we must improve our innovative performance and a manifesto for making that happen. This manifesto shows how enterprises with essential behaviors, tools, and processes can liberate the genius of staff to create great value for their customers, enterprises, investors, staff, and society."

—CURTIS R. CARLSON,
CEO, Practice of Innovation, LLC; former CEO of
SRI international (1998–2014)

"In *The Age of Agile*, Steve Denning—one of our leading management thinkers—demonstrates why and how a passion for products, not profits, is the key to successful business enterprise."

—WILLIAM LAZONICK,
Professor of Economics at the University of Massachusetts Lowell,
and codirector of its Center for Industrial Competitiveness

"In Silicon Valley, companies that grow big usually start to lose their way. Steve Denning, in *The Age of Agile*, shows how to avoid the common pitfalls start-ups make as they scale into larger, more mature companies. It is a blueprint for startups, small businesses, and large publicly traded companies alike."

—**STEPHEN FORTE**,
Managing Partner, Fresco Capital

"*The Age of Agile* should be a key companion to guide business leaders in their quest to address the biggest challenge set out by Peter Drucker for management in the twenty-first century: achieving a step change in the productivity of knowledge work. Without trying to give recipes, the book provides many lively and recent case studies and relevant background perspectives that will help leaders to re-frame the practice of current industrial-age management and to chart their unique way into an increasingly complex and unpredictable future."

—**RICHARD STRAUB**,
founder and President of the Global Peter Drucker Forum

"*The Age of Agile* is a tour de force. As a CEO strategy advisor, this is the most exciting book I have read this year. Finally, someone has laid out the core principles and, with the help of pertinent, easy-to-understand examples, lays out what it takes to apply Agile to any organization. Thank you, Steve Denning!"

—**SETH KAHAN**,
author of *Getting Change Right* and *Getting Innovation Right*

THE
AGE
OF

AGILE

How Smart Companies Are
Transforming the Way Work Gets Done

STEPHEN DENNING

**HARPERCOLLINS
LEADERSHIP**

AN IMPRINT OF HARPERCOLLINS

ISBN 978-1-4002-4240-5 (TP)

To Stephanie, for all that she inspires in me.

CONTENTS

FOREWORD

Bureaucracy is the default operating system for virtually every large-scale organization on the planet. Founded on the ideology of controlism, it elevates conformity above all other organizational virtues. Bureaucracy constitutes an organizational caste system that differentiates between the thinkers (managers) and the doers (employees). As a consequence, bureaucracy squanders enormous quantities of human initiative and imagination.

Today's creative economy needs a radical rethink of our top-down, tradition-encrusted management principles and processes. The challenge: building organizations that are as innovative as they are efficient, as passion-filled as they are pragmatic.

This is not merely about implementing a new practice, process, or structure. Instead, we have to start with a new set of management principles, among which four are particularly vital: transparency, competence, localization, and upside.

Nucor, the most consistently profitable steel company in the world, practices radical transparency. Every single associate knows the profitability of every order that ships. At Nucor, it is the frontline employees, not managers, who are responsible for maximizing margins.

Empowerment also depends on competence. Take Morningstar, the world's largest tomato processor. It has no managers and all key investment decisions are taken by individuals who in other organizations would be regarded as "blue collar" employees. Most of these employees are capable of sophisticated financial modeling—they can calculate the net present value and internal rate of return of new investments. Instead of moving decisions upward at Morningstar, they have moved competence down—to individuals who have the information and the context to make the best decisions.

If you want individuals to think and behave like owners, the organization must be decomposed into small, localized units—each with its own profit and loss (P&L) responsibility. All employees need to be able to see a direct line between their contribution and the profitability of the microbusiness in which they work. In large, monolithic organizations, there is little sense of accountability for results and little discretion in decision making. The result: employees who show up at work physically, but leave most of their discretionary energy at home.

Finally, if you want people to give their best, they must have an upside—that is, the possibility of personal and financial growth. In most organizations, fixed salaries and role definitions give individuals little incentive to do more than merely meet their targets. There is virtually no room for internal entrepreneurship.

These findings are entirely consistent with what Steve has observed in organizations that are committed to small, agile teams. Make no mistake, the management revolution is well under way. The only question for your organization is whether it is going to lead or follow.

—*Gary Hamel*
Professor, London Business School and
Director, Management Lab

INTRODUCTION

An unstoppable revolution is now under way in our society, affecting almost everyone. The revolution isn't being launched by opposition political parties, or by terrorists in secret cells, or through espionage by some obscure government department. The revolution is being conducted in plain sight by some of our largest and most respected corporations. It's visible to anyone with eyes to see. It's a revolution in how organizations are being run.

The revolution is very simple. Today, organizations are connecting everyone and everything, everywhere, all the time. They are becoming capable of delivering instant, intimate, frictionless value on a large scale. They are creating a world in which people, insights, and money interact quickly, easily, and cheaply. For some, the revolution is uplifting and beautiful. For others, it is dark and threatening.[1]

Dazzling examples of the new way of running organizations are everywhere apparent. Firms like Apple and Samsung offer devices that can be tailored to meet the individual wants and whims of hundreds of millions of users. Firms like Tesla, Saab, and Ericsson are upgrading cars, planes, and networks, not by physically installing new items, but by delivering new software to the products via the Web. Meanwhile, Spotify matches billions of musical playlists to individual users' tastes and delivers a weekly playlist tailored to each user's preferences, while Warby Parker sells high-quality eyeglasses for a small fraction of what traditional retailers charge by using a low-friction online model. Online services like Skype, Zoom, and WhatsApp are taking tens of billions of dollars away from old-guard telecom firms by giving customers free or low-cost calls. Amazon has demonstrated what can be accomplished when customer value is pursued ahead of short-term profits: It's not just the world's biggest retailer—it's bigger than all the other retailers put together.[2] Google has become big and rich very quickly, by providing search capabilities that are offered free.[3] The population of Facebook is bigger than that of

China. Airbnb, Uber, and Lyft are showing how to unlock the value in existing assets that were previously lying idle. And so on.

At the same time, what is lifting some companies is killing others. The examples here are also abundant. "Market-leading companies," as analyst Alan Murray has written in the *Wall Street Journal*, "have missed game-changing transformations in industry after industry—computers (mainframes to PCs), telephony (landline to mobile), photography (film to digital), stock markets (floor to online)—not because of 'bad' management, but because they followed the dictates of 'good' management."[4] In effect, the "good management" that these firms were practicing had become anachronistic. It simply didn't work anymore.

Spoiler alert. The difference between winners and losers *isn't* a matter of access to *technology* or *big data*. Both the successful and the unsuccessful firms generally have access to the same technology and data, which are now largely commodities. Traditionally managed organizations also use digital technology and big data but typically get meager results. In some cases, like Kodak, it's the firm that invented the new technology that has failed to exploit it. It's not access to technology and data that makes the difference. The difference lies in a different way of running the organization that deploys technology and data more nimbly.[5]

Trying to exploit technology and data with the management practices that are still pervasive in many big corporations today is like driving a horse and buggy on the freeway. To prosper in the very different world that is emerging, firms need a radically different kind of management.

Some firms are embracing the new management paradigm with alacrity. They are happy to shed the traditional management practices of manipulating both staff and customers and instead follow their natural preference to treat people as people and engage in authentic adult-to-adult conversations. Some of them are generating inspired workplaces that create meaning in people's lives.

Other firms are getting on board more gradually. They reflect on the obvious anomalies of traditional management and feel frustrated that their efforts to fix things don't work. They find themselves having to run faster and faster just to stay in place. Yet they can also see the extraordinary gains of firms operating in the new way and begin to wonder: "Why can't we have what they are having?" There often follows a lengthy period of reflection and experimentation before managers finally "get it" and internalize the new mindset.

Some firms are actively resisting the change. For established organizations that have been successfully managed in a traditional fashion for many years with settled processes, routines, attitudes, and values, the new management paradigm can be difficult, even baffling. It is often at odds with the unspoken assumptions about "the way we do things around here."

Still other firms have sought to avoid the dilemma through financial engineering. They are pursuing ways of extracting value from the corporation through short-term cost-cutting, offshoring, share buybacks, tax gadgets, and other devices. While these expedients can create an appearance of prosperity for investors in the stock market, they are systematically destroying real shareholder value and genuine economic well-being.

When managers do embrace the new way of running the organization and the "Aha!" of how the new management paradigm is happening on a large scale, it can be an amazing and humanizing experience. Why would anyone consider doing things differently?

In short, this book explores how some organizations are learning how to operate in a way that is potentially better for those doing the work, better for those for whom the work is done, better for the organizations themselves, and better for society. These organizations now form a vast global movement that is transforming the world of work.[6]

The movement began many decades ago, but took off in a major way more recently in an unexpected place: software development. It is now spreading rapidly to all parts, and all kinds, of organizations—big firms and small, simple and complex, software and hardware, technology, manufacturing, health, pharmaceuticals, telecommunications, aircraft, and automobiles—you name it.

The new paradigm enables organizations to thrive in a world of rapid and unpredictable change. It enables a team, a unit, or an entire enterprise to nimbly adapt and upgrade products and services to meet rapidly changing technology and customer needs with efficiency gains, quality improvements, or even completely new products and services. It permits an organization to flourish in a marketplace that is increasingly volatile, uncertain, complex, and ambiguous—the so-called VUCA world.

■

How did this revolution happen? Some, but not all, of the organizations implementing the new management paradigm see the foundational document of the movement as the 2001 *Manifesto for Software Development*—now commonly called the Agile Manifesto. Others refer to earlier historical antecedents and management practices and use terms like "lean," "quality," "design thinking," or their own home-grown label.[7]

The Agile Manifesto declared that "uncovering better ways of developing software" requires a reversal of some fundamental assumptions of twentieth-century management. It values "individuals and interactions over processes and tools, working software over comprehensive documentation, customer collaboration over contract negotiation and responding to change over following a plan."[8]

Yet in propounding these values, the Manifesto was implicitly raising a wider and deeper set of questions. What if firms could create workplaces that drew on all the talents of those doing the work? What if those talents were totally focused on delivering extraordinary value to the customers and other stakeholders for whom the work is being done? What if those receiving this unique value would be willing to offer generous recompense for it? What would these workplaces look like? How would they operate? How would they be reconciled with existing goals, principles, and values? Could they operate on a large scale? Could they be reliable?

In 2001, no one really knew the answers to those questions. Experiments were conducted to find out. As with anything new, things proceeded in fits and starts, with frequent setbacks. Many variations in practices were explored. Even when the practices were in essence the same, the approaches often had different labels.

The initial experiments were with single teams. As some of these experiments succeeded, the experiments expanded to groups of teams and eventually to large-scale implementations, even whole organizations. The new way of running organizations spread to manufacturing and other fields.[9] Some startups that began operating in an Agile fashion continued to be run this way, even as they grew.

For some years, it was hard to make sense of what was going on. Even some of those who embraced the new management paradigm saw it as playing a limited role, mainly in simple software activities in small units or in organizations where reliability was not an issue. Many teams and firms that claimed to be operating in the new way were doing so in name only. Some suggested that the new way of managing, as it expanded beyond individual software development teams, would inevitably mutate into the traditional practices of top-down bureaucracy in order to achieve efficient, reliable management in large-scale operations.

Yet over time what was working and what wasn't became apparent. As a result, there was a convergence toward a family of goals, principles, and values that is demonstrably more productive and more responsive to today's marketplace than traditional management and that can operate on a large scale. As the movement matures, and as managing software becomes central to the success of most businesses, the new paradigm is becoming a key to the management of everything.

The new paradigm has not been easy for traditional managers to understand or implement. First, much of the recent momentum came from an unexpected source: software development, which had no prior reputation for excellence in management. It was hard for general managers to accept that they had anything to learn about management from software developers. Managers were slow to grasp the wider significance. In some ways, the new way of running an organization is still the best-kept management secret on the planet.[10]

The antagonism is understandable. These managers have spent most of their careers accepting the prevailing management paradigm and proceeding within its assumptions. Their careers have flourished by mastering and implementing twentieth-century concepts and practices. They see that business schools still teach these concepts and practices. The thought that everything on which they have built their careers is changing beneath their feet can be unnerving—even alarming. Yet the change is coming at them willy-nilly. One study suggests that 75 percent of the Standard & Poor's 500 Index (S&P 500) will turn over in the next fifteen years.[11] Another says that one in three public companies will delist in the next five years.[12] The choice for many organizations is simple: change or die.

Second, the illusion that technology will by itself solve the challenge of adaptability is still widespread. Many firms fail to see that since generally all organizations have access to the same rapidly evolving technology, competitive advantage flows not from the technology itself but rather from the agility with which organizations understand and adapt the technology to meet customers' real needs.

And third, the way of running an organization represents a genuine paradigm shift in management, with fundamentally different goals, principles, and values that disrupt deeply entrenched assumptions, attitudes, and habits. Traditional managers often believe—and hope—that the changes are merely a fix that they can apply to specific issues, rather than a fundamentally different way of approaching management itself.

The revolution is proceeding at different speeds in different sectors. Manufacturing, for instance, which pioneered the early stages of the Agile revolution, is now behind software development. Yet as physical products and services are increasingly software driven and the "Internet of Things" makes its presence felt, the distinction between software and manufacturing is disintegrating. As "software is eating the world," all firms are becoming dependent on software, thus accelerating the spread of the Agile paradigm.[13]

The central theme of this book—that corporations must radically reinvent how they are organized and led and embrace a new management paradigm—may seem to some readers to be extreme. It is not. This isn't a management fad that was invented last Tuesday and will be gone by Friday. It is based not just on a handful of recent examples—mere flashes in the pan—but on the experiences over decades of tens of thousands of organizations around the world.

As someone who had been deeply involved in general management for decades as a manager at the World Bank, I have to admit that I didn't pay much attention to these developments until 2008. It was only then that I suddenly grasped that the management discoveries of these software developers had vast implications for all organizations. I introduced the thinking to general managers in my 2010 book, *The Leader's Guide to Radical Management*. Since then, I have been studying the rapidly expanding implications in many different sectors, as whole organizations join the movement. I have written over 700 articles as a contributor to *Forbes.com*, with many case studies—both recent and historical.

For the last couple of years, I have been leading a learning consortium of major organizations that are passionate about discovering together what these changes mean for their goals, principles, and practices. This book is in part a progress report on their discoveries.[14]

Implementing the new management paradigm isn't easy. It's not for the faint of heart. All the firms that we studied experienced major setbacks in the early going. The leaders persevered and eventually succeeded through adherence to their goals, principles, and values.

Nor should we be distracted by the fact that a great deal of fake change is still rampant. In some cases, organizations claim to be operating in the new way while offering no more than a thin veneer laid on top of traditional top-down bureaucracy. These companies are doing what they've always done; they're just giving it a new name.

The new management paradigm is a journey, not a destination. It involves never-ending innovation, *both* in terms of the specific innovations that the organization generates for the customer *and* the steady improvements to the practice of management itself. A firm never "arrives" at a steady state where it can relax because "we are now Agile." Embracing the new paradigm requires continuous commitment and leadership from management.

This book offers snapshots of firms that are at different stages of their respective journeys. What got them to where they are now is no guarantee of future success. These firms will only continue to prosper if they persist in their embrace of the new goals, principles, and values and go on delighting customers with continuous innovation.

This book does not of course begin with a clean slate. There is a vast and growing body of literature on the new management goals, principles, and practices. Particularly practices. Much of the literature is written by software developers for other software developers, often in software-centric jargon and often focused on tools and processes. This book distills the essence of the new management paradigm, particularly the relevant mindset, in nontechnical language.

The first part of the book (Chapters 1 through 7) covers the principles of Agile management. To master the heart of the new management paradigm, we will begin in Chapter 1 by visiting two very different firms—a very young firm, Spotify, the music streaming service, and a very old firm, Barclays, the global bank—with one key thing in common: a ferocious commitment to the new management paradigm.

Then we'll explore the three laws of the new management paradigm: the Law of the Small Team, the Law of the Customer, and the Law of the Network. We will see how the new management mindset applies the three laws. This isn't just a methodology or process to be implemented within the assumptions of current management practice. It involves a fundamentally different concept of what an organization is and how it must operate to succeed in today's marketplace.

We'll begin with the *Law of the Small Team* (Chapter 2) because it's the aspect of management that received most of the attention of the early Agile implementations. In our visits to Menlo Innovations (the developer of mission-critical software in Ann Arbor, Michigan) and Etsy (the handicrafts marketplace), we'll see how, in a VUCA world, big and difficult problems need to be disaggregated into small batches and performed by small cross-functional autonomous teams, working iteratively in short cycles in a state of flow, with fast feedback from customers and end-users.

We then take a look at the *Law of the Customer* (Chapter 3). It's the most important of the three laws because it makes sense of the other two. We'll examine the implications of the epic shift in power in the marketplace from seller to buyer, and the need for firms to radically accelerate their ability to make decisions and change direction in the light of unexpected events.

Then we'll take a tour of the lynchpin of Agile management—the *Law of the Network*—which ties together the other two laws. We'll see (in Chapter 4) what's involved in making the whole organization Agile. We'll learn how even the U.S. Army has discovered that a steep, vertical hierarchy is no match for a committed, interactive network, even one that is underresourced and underskilled. We'll probe yet more Agile paradoxes: Control is enhanced by letting go of control, and Agile leaders are less like fierce, conquering warriors than curators or gardeners.

We'll pay a visit to an old global behemoth in Seattle that, against all the odds and the predictions of Agile experts, is undergoing an *Agile transformation at scale* (Chapter 5). Then we'll explore what's involved in making major financial gains by moving from operational Agility to *Strategic Agility* (Chapter 6).

Achieving Strategic Agility will often involve a *shift in an organization's culture* (Chapter 7). We'll pay a call on a Silicon Valley icon that was on the verge of bankruptcy and that turned its toxic culture into a dynamic innovative culture that is now generating huge profits through market-creating innovations.

In the second part of the book (Chapters 8 through 11), we'll explore key constraints or traps to implementing Agile management. First up is the pervasive goal in publicly owned corporations: *maximizing shareholder value* as reflected in the current stock price. We'll see the havoc caused to Agile management and to the economy by managers who focus on extracting, rather than creating, value (Chapter 8).

We'll look at the mind-boggling use of *share buybacks* and examine what organizations—and society—need to do about it (Chapter 9). We'll examine the problems caused by *cost-oriented economics* and the resulting large-scale offshoring of jobs over the last several decades (Chapter 10). We'll see how a backward-looking strategy became a constraint instead of an enabler (Chapter 11).

In the epilogue (Chapter 12), we'll explore the historical precedents over four centuries for making paradigm shifts in management and the leadership implications for the emerging age of Agile, which offers the possibility of a great awakening—the foreshadowing of a transformation in the way our organizations and our society function.[15]

This book answers three simple questions. How do organizations flourish in a VUCA world, where the customer is in charge of the marketplace? Why has embracing this new way of running organizations become a necessity? What can leaders at all levels of the society do to create a more energizing, prosperous, and meaningful mode of working and living?

[PART ONE]

AGILE
MANAGEMENT

1.

MORE VALUE FROM LESS WORK

What we need is an entrepreneurial society in which innovation and entrepreneurship are normal, steady, and continuous.

—PETER DRUCKER[1]

For Spotify—the fast-growing Swedish music streaming service with over 100 million active users and more than 30 million paying users—the true value of Agile management became dramatically apparent in mid-2015. Spotify had embraced Agile management since its launch in 2008, with swarms of self-organizing teams intent on delivering steadily more value to Spotify's users. The premise of Agile management is that empowering bottom-up innovation will steadily add significant value for customers and the firm. Accordingly, the teams at Spotify—some 2,500 people as of mid-2016—seek to learn everything about you as both a listener of music and as a user of Spotify, and then find interesting ways to appeal to you on both levels. Sometimes they do that by matchmaking, by telling you at the right moment about a great playlist, or a great new feature, or some new content that they think you will like. Other times, they do it by creating new listening experiences.

Innovations generated by Agile teams had fueled Spotify's growth for seven years. But in March 2015, a couple of Spotify's software engineers—Chris Johnson and Ed Newett—came to Matt Ogle, a senior product leader with two degrees in English literature and a background as an engineer, with an idea that turned out to be a game-changer. They had thought of a way of solving a problem that had stumped Spotify and other music streaming services like Pandora and Apple Music for years: How could users find the music they would really love in a library of millions of songs without wasting time browsing through music they didn't like?[2]

In 2013, Spotify had introduced a feature called News-Feed in which users received personalized recommendations of albums and artists. This was progress, but it still took a lot of effort on the part of users to engage with the recommendations and get to listen to the music.

In 2014, Spotify had offered a feature called Discover, which grouped the recommendations into strips, as on Netflix. This was easier to use than News-Feed, but it also required active user effort. Studies showed that users were still spending more time listening to playlists that Spotify's editors had created.

Now the two engineers had another idea. What if, they asked, we could completely remove the friction for you as a user? What if we took the music you had listened to in the past and sorted it into micro-genres? What if we analyzed the billions of playlists created by other users and algorithmically matched preferences with your playlists, so we can then create a new playlist specifically designed for you? What if we delivered this personalized playlist of songs for you once a week? What if, every time you skipped a track, we learned from that and made sure that your next weekly playlist would appeal to you even more? What if we did this not just for you but for every one of our tens of millions of active users? Would that be possible? Or would it just produce noise? This was the embryo of the wildly successful idea on Spotify that became known as Discover Weekly.

Ogle liked the idea. He discussed with the engineers different ways of making it work. They brought in a designer who played the bad cop in the discussion. "Why should this feature exist?" he asked. "We already have too many things for users! What will it do that we're not already doing?" Those questions helped the team get clearer on what the new idea was for and what value it might add.

Ogle's team had all the elements in place to conduct a quick experiment. Spotify had already collected data on active users—which then numbered some 75 million. They had also built high-level capabilities in machine learning and artificial intelligence. They had already developed micro-genres of music and classified its entire vast repository of music and its billions of playlists.

But most important, Spotify had created an organizational culture of Agile management in which autonomous cross-functional teams were encouraged to experiment and create new ways of adding value to customers. With Agile management, Ogle and his team didn't need to prepare a detailed cost-benefit proposal and seek a series of approvals up a steep management chain before they could try out their idea. They were used to working as a team, with radical transparency among the team members. They were already tightly focused on the user experience: They knew how to test alternatives and learn from the tests. Within a couple of weeks, the tiny cross-functional team had pulled together a quick prototype and tried it out on Spotify's own staff—all active Spotify users.

The result? Spotify staff just loved it. Ogle himself became a huge enthusiast. On one of his very first playlists, he recalls listening to a song by Jan Hammer, the composer of the *Miami Vice* theme. "It starts off with this poppy thing, then the strings," said Ogle. "When the vocals came in, I thought, holy shit, we have to ship this feature. Whatever just served this song needs to be out in the world."[3]

Ogle and his team did another quick experiment on one percent of the active Spotify users—close to a million people. Again, the response was strongly positive. Amazingly, 65 percent of respondents found "a new favorite song" in their personalized weekly playlist. As a result, Spotify's management was ready to introduce Discover Weekly for all Spotify listeners.

Scaling up the Discover Weekly algorithms from one million users to 75 million users in twenty-one languages in multiple time zones each week proved to be more of a challenge than the engineers expected. Nevertheless, working in an Agile fashion totally focused on the goal, the team took only a couple of months to complete the work. When Discover Weekly was deployed to all Spotify users in July 2015, just four months from the initial concept, it was a wild success—beyond anything Ogle and his engineers had imagined.

In fact, Discover Weekly has become a global phenomenon. It has resulted in a massive boost for Spotify's brand and a huge influx of new users. It is more than just another feature—it is almost a new brand in itself, with foreign-language countries clamoring for "Discover Weekly" rather than a label in their own language. Every Monday morning, Spotify users—now more than 100 million of them—receive a playlist of thirty songs that feels like a gift from a talented and knowledgeable musical friend who understands their taste in music and who has searched the world to put together a handpicked playlist of the very best music they will adore.

Users say it's spooky how fresh and familiar their Discover Weekly playlists feel. A common reaction is, "How come Discover Weekly knows me better than myself?" Within the first six months, songs from Discover Weekly had been streamed several billion times.

"If you're the smallest, strangest musician in the world, doing something that only twenty people in the world will dig," says Ogle, "we can now find those twenty people and connect the dots between the artist and listeners. Discover Weekly is a compelling new way to do that at a scale that's never been done before."[4]

Discover Weekly gives Spotify a massive brand advantage over competitors like Pandora and Apple Music, which also have vast catalogs of music but without Spotify's personalized approach to help you find music you will enjoy. Yet Spotify knows it can't rest on this success. It knows that its competitors will soon emulate Discover Weekly. In the spirit of Agile, Spotify is already racing ahead with further innovations that will bind its user community ever more tightly to the music streaming service they have come to love. Spotify's management knows that it will only survive if it continues to pursue Agile management and innovates faster than its competitors.

■

At first glance, the idea that Barclays—a 327-year-old transatlantic bank with more than 100,000 employees—could become as Agile as Spotify and deliver an instant, frictionless, intimate banking experience at scale might seem ridiculous. The bank operates in a difficult environment. It's highly regulated. It's recovering from a major financial crisis. And it has new challenges coming its way as it grapples with what Brexit means for the future. It's a transatlantic bank offering products and services across personal, corporate, and investment banking, credit cards, and wealth management, with a strong presence in its two home markets: the United Kingdom and the United States. The bank operates in over forty countries. One-third of payments made in the U.K. pass through Barclays.

Despite its size and reach, Barclays, like all the big global banks, finds itself in a world in which its customers are coming to expect the same kind of instant, intimate, frictionless responsiveness at scale that they experience with Spotify's Discover Weekly playlist. What they would like from a bank are prompt, helpful responses, not just to simple questions like, "What's my bank balance?" but also, "Should I spend this money on this car? Should I buy, lease, or get a loan? What sort of insurance do I need? What impact will this have on my savings? How is my retirement looking?"

Ironically, if Barclays, like all banks, knew what it already knew, it could, like Spotify, provide good answers to those questions. That's because Barclays, like Spotify, has massive amounts of data about its customers and clients that could help the bank add value by answering some of these questions—if this data was put to better use.

Barclays, like all the big global banks, also faces the challenge of inherited management practices and old data structures and processes that can make it difficult to deliver instant, frictionless, intimate banking experiences at scale. As an old institution, it has to invest time and money in upgrading and streamlining old data structures, processes, and applications to improve the products and services it can offer its customers. With individual customers' information stored on separate parts of the bank's systems, it can be difficult to provide the level of customer experience that customers have come to expect from companies today.

Decades ago, none of that mattered to customers when they would make an appointment to see their local bank manager and discuss how they could finance an extension on their house, or buy a new car, or invest in shares to provide for their retirement. Much depended on a personal relationship with the bank manager. In the end, customers might have gotten the loan or the advice they needed. But the process took time to arrange. It was inconvenient. It involved significant effort producing documents and records. And it worked for only a handful of customers.

How could Barclays possibly go from that old-school banking experience to the kind of easy, convenient, personalized responsiveness that its millions of twenty-first-century customers are coming to expect in everything they do?

Unlike Spotify, Barclays had not, until recently, made an institutional commitment to Agile management. Like most of the big global banks, Barclays had management practices and processes that were steeply hierarchical and bureaucratic. As in most large organizations today, Barclays had isolated islands of Agile software developers, but this wasn't consistent across the organization. Nor had Barclays developed the capability to carry out and learn from rapid experiments with its customers.

Yet management could see that some of the bank's competitors are already acting with that kind of agility. On the international front, large Asian competitors, such as Alibaba and Tencent, are expanding rapidly, with management and data systems oriented to meeting their customers' needs, not their own internal requirements. They are acquiring new customers for just a couple of cents each, while the cost for larger, mainstream financial players like Barclays was a large multiple of that. Meanwhile, digital behemoths like Google, Amazon, and Apple are handling an increasing proportion of the world's financial transactions that used to be handled by banks. On yet another front, financial sector startups are innovating rapidly and threatening to create the very instant, personalized, frictionless responsiveness that customers increasingly expect. Although these startups are still small, many of them are now valued at more than a billion dollars. A group of some thirty of them formed in just the last few years are worth collectively more than Barclays' own market capitalization.[5]

In 2014, Barclays started to respond to these external threats, recognizing that it needed to adapt its way of working, management practices, and systems. To keep pace with what new competitors were offering, the bank recognized that it would only succeed if it was able to offer, like Spotify, easy, quick, convenient, personalized responsiveness at scale. In short, Barclays had to become Agile. In March 2015, Barclays' operations and technology team announced that becoming Agile was a key strategic initiative. The many islands of Agile within Barclays were invited to come out from the shadows and become the champions of Agile transformation at Barclays.

Two years later, by March 2017, Barclays had made remarkable progress. After a massive program of Agile training and coaching, the equivalent of more than a thousand self-organizing Agile teams—around 15,000 people—were operating, covering every aspect of the business (commercial bank, investment bank, accounts, audits, and compliance)—all focused on delivering more value sooner to customers. While some teams were still in the early stages of mastering Agile management practices, a sizable proportion were already mature and making significant inroads improving Barclays' performance.

For example, the *Agile Loans* team devised an online loan application that reduced the time from "initial ask" to "the Barclays response" from twelve days to twenty minutes. The online solution operates almost a thousand times faster than cumbersome paper-based procedures and in-person interactions requiring visits to physical offices.

Agile Onboarding is another example. You might think that a bank like Barclays would have figured out how to make it easy for anyone to become a customer, either as an individual or as a business, through opening a new account, applying for a loan, or making a new investment. Yet in practice, the online and in-branch processes for this function were cumbersome. Barclays had recognized the importance of improving the process: Several years before, it had tasked a group of over a hundred developers to create a better online onboarding process, but they had yet to produce a desirable solution. By contrast, when Barclays put together an Agile team of only six developers, they developed a viable onboarding tool in just six months, also resulting in cost savings and improved productivity.

The Agile transformation at Barclays is both top-down and bottom-up. Support and inspiration from the top are critical to legitimizing the Agile way of working and helping to remove impediments to the fundamental changes implicit in Agile. At the staff level, there is a community of Agile practitioners with some 2,500 people, along with around 15,000 practitioners involved in Agile teams. In fact, it was these Agile community members who drafted the bank's Agile Enterprise transformation proposal in 2014. Barclays' own Agile advocates said, "Here's what Agile means for us and here's how we should do it." And now Barclays is doing exactly that. Since Agile is perceived as something inspired and supported—not imposed—from on high, the psychological level of buy-in at the working level is strong.

Nevertheless, it's still early days for Barclays' Agile transformation. Barclays has much work to do before it can fully deploy the intelligence and the know-how of algorithms that could help millions of customers make better financial decisions in real time. This will require deeper understanding of their different situations, contexts, and prospects, as well as insight into how their needs can be met on the fly and at scale.

And it is only through years of steady implementation that Barclays will be able to claim that it has embedded a culture of Agility throughout the organization.

But Barclays has made a start. And everywhere there is evidence of early success. Its management knows that Agile transformations involve a deep change in organizational culture that will take time. Barclays can take heart from the experience of other global behemoths like Microsoft and Ericsson that have persevered with Agile transformations at scale over a period of years and have made sustained progress. In other words, firms don't have to be "born Agile," like Spotify. Even big, old firms can undertake an Agile transformation if they set their minds and hearts to it—and stick with it.

Spotify and Barclays are just two of thousands of organizations that have made a fundamental discovery: Business success in the twenty-first century increasingly depends on becoming as nimble as the rapidly shifting and unpredictable context in which they find themselves.

As a result, Agile management is now spreading to every kind of organization and every aspect of work, as recognized in 2016 by the citadel of general management, *Harvard Business Review*, with its article "Embracing Agile," by Darrell K. Rigby, Jeff Sutherland, and Hirotaka Takeuchi.

> Now agile methodologies—which involve new values, principles, practices, and benefits and are a radical alternative to command-and-control-style management—are spreading across a broad range of industries and functions and even into the C-suite. National Public Radio employs agile methods to create new programming. John Deere uses them to develop new machines, and Saab to produce new fighter jets. Intronis, a leader in cloud backup services, uses them in marketing. C. H. Robinson, a global third-party logistics provider, applies them in human resources. Mission Bell Winery uses them for everything from wine production to warehousing to running its senior leadership group.[6]

The centrality of Agile for general management was also recognized in March 2016 with McKinsey & Company hosting a Global Agility

Hackathon involving some 1,500 participants worldwide. According to the winning submission:

> Becoming an agile organization is an increasingly urgent necessity for companies in today's digital economy, yet most companies have a deeply embedded command organization architecture and culture. This reflects, first and foremost, the industrial economy mindsets and skills of their senior leaders, which is arguably the greatest obstacle to becoming an agile organization. . . . To make the transformation, senior leaders must learn and practice a holistic and complete set of new mindsets and skills, and apply them to design a wholly new, agile organization architecture and culture.[7]

The emergence of Agile for general management was also confirmed by the findings of the SD Learning Consortium, a group of companies including Microsoft, Ericsson, CH Robinson, Riot Games, Barclays, and Cerner. Amid conflicting claims as to whether Agile is something real, the Consortium seeks to discover what is actually happening on the ground and to learn what works and what doesn't. Site visits showed that in a few cases, Agile was indeed hardly more than talk; the organization was still functioning as a top-down bureaucracy. But in most cases, site visits showed that major corporations have large-scale implementations of Agile goals, principles, and values. In effect, Agile management in large organizations is actually happening.[8]

In organizations that have embraced Agile management, self-organizing teams are continuously providing new value for customers. Because the work is done in an iterative fashion with continuous interaction with users, the organization can constantly upgrade what it offers to users, sometimes in real time. When teams work in a common cadence—short iterations of the same length that start on the same day—many teams can work together on large, complex challenges in a coordinated fashion and deliver fail-safe products. When Agile is done right, the teams are working within a business model in which the organization is generating value for the organization itself as well as the customer. Everything—the work being done, the information, and the money—moves easily, in an integrated fashion, often with strong returns to scale.[9]

Agile management is about working smarter rather than harder. It's not about doing *more work* in less time: It's about generating *more value* from *less* work. For instance, in 2011, Ericsson (a 140-year-old Swedish firm with around 100,000 employees) embraced Agile for its business unit in managing networks for the world's telecommunications companies. Competition in the sector is fierce. Of seven global firms that operated just a few years ago, only three, including Ericsson, have survived. Before 2011, Ericsson would build its systems on a five-year cycle with a unit housing several thousand employees. When the system was finally built, it would be shipped to telecom customers and there would be an extended period of adjustment as the system was adapted to fit their needs.

Now Ericsson has over a hundred small teams working with its specific customers' needs instead of working on all of the requirements for the overall system. They involve the customers in testing different aspects of the system in three-week cycles. The result is faster development that is more relevant to the specific needs of the customers. Moreover, the interaction has enabled Ericsson to focus on the customers' highest priorities. The client gets to see the next iteration of the system every three weeks, instead of waiting five years for one "big bang" delivery. They know exactly what is being worked on and can direct the course of the work.

In one case, Ericsson's client said, "For us to deploy this new system in our entire network, we would need around 120 improvements." It turned out, because of Ericsson working interactively with its customers, the client decided to go ahead with around sixty of the improvements. The client said: "If we had been working as we were in the past, we would never have done that at this stage. But because of the cooperation that we have, now we are ready to go ahead." The result? The client gets value sooner. Ericsson has less work in progress. And Ericsson is deploying one to two years earlier than it otherwise would, so its revenue comes in one to two years sooner. The client is much happier and there is a financial benefit for Ericsson.[10]

One common misunderstanding of Agile management is to see it as a technology solution—digitization—rather than an approach to running an organization. While obviously Agile does use digital technology to enable instant, frictionless, intimate value at scale, the gains are driven by Agile *management*. When top-down bureaucracies use digital technology, machine learning, platforms, blockchain technology, or the Internet of Things, they typically get meager results. That's in part because internally driven innovation often generates changes that customers don't want or aren't willing to pay for.

Resolving complex problems requires both continuous collaboration across internal silos and interaction with customers. Delivering instant, intimate, frictionless customer experiences that delight customers lies beyond the performance capability of an internally focused bureaucracy. Bureaucracy was never intended to accomplish that: It was designed to produce merely consistent average performance according to internal rules set by the organization itself.

Moreover, bureaucracies, with their steep chains of command, can't move fast enough to take advantage of opportunities in a VUCA marketplace as they emerge. In a competitive setting, it's not technology itself that makes the difference. The key is how adroitly the firm uses the technology. The driver of sustained success is the Agile mindset.

But what exactly is Agile? What does it mean for an organization to embrace the Agile mindset? When I use the word "agile" or "nimble," you might think about a squirrel or a ballet dancer or a champion soccer player. You probably wouldn't think of a large organization—big, unwieldy, clunky, slow, out to make money from you, and fundamentally unfriendly. You generally don't think of organizations as being nimble because generally they're not. We're accustomed to dealing with organizations that are frustratingly set in their ways and preoccupied with their own internal processes. Their motto could be: "You take what we make; that's the way it is." The possibility that organizations could operate with agility is thus not obvious. And yet the site visits of the SD Learning Consortium show that large Agile organizations are not only possible. They do exist.

The challenge of understanding Agile is magnified by the more than forty labels that have been used for different flavors of Agile.[11] Nor does the multiplicity of Agile practices help: In my 2010 book, *The Leader's Guide to Radical Management*, I identified more than seventy different Agile practices. How on earth are traditional managers going to make sense of such a bewildering blizzard of ideas?

The truth is that when we look closely, we can see that organizations that have embraced Agile have three core characteristics:

1. The Law of the Small Team
2. The Law of the Customer
3. The Law of the Network

The first universal characteristic of Agile organizations is the *Law of the Small Team*. Agile practitioners share a mindset that work should, in principle, be done in small, autonomous, cross-functional teams working in short cycles on relatively small tasks and getting continuous feedback from the ultimate customer or end-user. Big and complex problems are resolved by descaling them into tiny, manageable pieces.

For the first decade of the Agile movement (i.e., the 2000s), much of the effort was spent on figuring out how to generate high-performance teams on a systematic basis. Teams of course were not a new idea. We all know the magic. We have all at one time or another been involved in a small group where communications flow effortlessly and the group seems to think and act as one. When we are members of such a team, we can analyze a situation, decide, and act as though we are part of a single, uninterrupted motion. There is no one in charge telling us what to do. We trust the other members of the team. That trust is rewarded by performance. It's as though the group has a mind of its own. Face-to-face conversation sorts out any differences in point of view. Work has become fun. The group is in a state of flow.[12]

Work in most twentieth-century organizations was very different. Big systems were implemented with big plans delivering large quantities of a standard product and succeeded through economies of scale. Work was broken down into small, often meaningless pieces. Individuals reported to bosses who ensured consistent and accurate performance in accordance with the plan's specifications. The boss's boss did the same, and so on, up the line. Plans and budgets were generated and allocated, division by division. The connection between any piece of work and its impact on the life of a customer was often hidden by mind-numbing internally focused systems. The result today is a disengaged workforce, with only one in five employees fully engaged in the work. Even worse, almost one in seven employees is actively disengaged and intent on deliberately undermining what the organization does.[13]

Throughout the twentieth century, writer after writer suggested that working in teams would be a better way to get work done. It began with Mary Parker Follett in the 1920s, and continued with Elton Mayo and Chester Barnard in the 1930s, Abraham Maslow in the 1940s, Douglas McGregor in the 1960s, Tom Peters and Robert Waterman in the 1980s, on to Douglas Smith and Jon Katzenbach in the 1990s and Richard Hackman in the 2000s.

Yet most organizations stayed stubbornly bureaucratic, with bosses supervising individuals. One reason was the pervasive management belief that teams couldn't deliver disciplined, efficient performance at scale: They were useful for solving complex one-off problems, but for the run-of-the-mill work in a big organization, the conventional wisdom was that bureaucracy was better.

Another reason was that most teams in twentieth-century organizations were teams in name only. Most of them weren't real teams at all. The team leader acted like any other boss in a bureaucracy. (See Figure 1-1.)

A third reason is that self-organizing teams achieving high performance were a rarity. The literature on teams often talked about high-performance teams—teams that were not just 10 percent or 20 percent better, but two, three, or even many times better—but suggested that they were a matter of luck. The right people had to have come together. The context had to be conducive. The stars had to be aligned. You could

Figure 1-1. Agile vs. bureaucratic team.

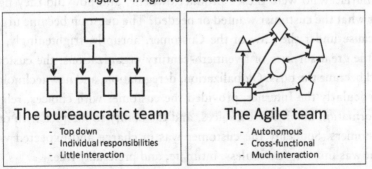

The bureaucratic team
- Top down
- Individual responsibilities
- Little interaction

The Agile team
- Autonomous
- Cross-functional
- Much interaction

create conditions where it *might* happen. You could encourage it. But ultimately a high-performance team was a rare accident.

It was the Agile movement that figured out how to create an environment that fostered consistent high performance in its teams. If there were a Nobel Prize for management, which there isn't, and if there were any justice in the world, which there isn't, the creators of Agile would be Nobel laureates. It is a breakthrough achievement, well accepted in the world of software development, and just now becoming more widely understood and recognized in general management.

■

The second characteristic of Agile organizations is the *Law of the Customer*. Agile practitioners are obsessed with delivering value to customers. The primary importance of the customer is recognized in the first principle of the Agile Manifesto:

1. Our highest priority is to satisfy the customer through early and continuous delivery of valuable software.

But frankly, in the first decade after the Agile Manifesto, customer focus received secondary consideration among software developers: Most of their attention was on getting the characteristics of the high-performance team right. In this period, teams often had very little contact with the actual customer. Instead, the customer was represented by a proxy representative who was mysteriously called a "product owner" and who supposedly knew what customers wanted.

Once Agile had solved the problem of how to generate high-performance teams on a consistent basis, then attention turned to the epic shift in power in the marketplace—the shift in power from seller to buyer. Who were these "product owners" and how did they figure out what the customer wanted or needed? The question became urgent, because under the Law of the Customer, abruptly, frighteningly, and to the great surprise of twentieth-century organizations, the customer had become the boss. Globalization, deregulation, and new technology, particularly the Internet, provided the customer with choices, reliable information about those choices, and the ability to interact with other customers. Suddenly the customer was in charge and expected value that was instant, frictionless, intimate, and preferably free.

As a result, firms had to think about the customer in a new way. Twentieth-century firms had gotten used to the notion that they could exploit and manipulate customers. If a customer didn't like something they were offering, the firm would say, "We hear what you're saying, but that's what we're offering. Take it or leave it. We'll consider introducing changes in our next model, now some years away." In today's competitive marketplace, this approach is steadily less effective. The customer is thinking: "Why are we waiting a couple of years? If you won't fix it now, I'll find someone who will."

The primacy of the customer is at once the most obvious and the most difficult aspect of Agile to grasp. One reason that it's difficult to understand is that twentieth-century managers had learned to parrot phrases like "The customer is number one!" while continuing to run the organization as an internally focused, top-down bureaucracy interested in delivering value to shareholders.

It's not that bureaucratic organizations ignore the customer. They do what they can for the customer—but only within the limits of their own internal systems and processes. Firms may say they are customer-focused, but if the information they need to answer simple questions from customers is hidden in multiple systems that don't talk to each other, or if customer services must be cut to meet a quarterly profit target, then it's too bad for the customer. The customer gets the short end of the stick. In a top-down bureaucracy, "the customer is number one" is just a slogan: Internal systems, processes, and goals take precedence. (See Figure 1-2.)

In the Agile organization, "customer focus" means something very different. In firms that have embraced Agile, everyone is passionately

Figure 1-2. Bureaucratic vs. Agile organization.

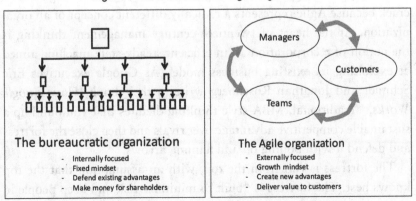

The bureaucratic organization
- Internally focused
- Fixed mindset
- Defend existing advantages
- Make money for shareholders

The Agile organization
- Externally focused
- Growth mindset
- Create new advantages
- Deliver value to customers

obsessed with delivering more value to customers. Everyone in the organization has a clear line of sight to the ultimate customer and can see how their work is adding value to that customer—or not. If their work isn't adding value to any customer or user, then an immediate question arises as to why the work is being done at all. The firm adjusts everything—goals, values, principles, processes, systems, practices, data structures, incentives—to generate continuous new value for customers and ruthlessly eliminates anything that doesn't contribute.

The third characteristic is the *Law of the Network*. Agile practitioners view the organization as a fluid and transparent network of players that are collaborating toward a common goal of delighting customers.

In the early years of the Agile movement, it was generally assumed that if the firm could get high-performance teams going, then the whole organization would be "Agile." It turned out not to be the case. It wasn't enough to have Agile teams totally focused on delivering more value to the customer if the rest of the organization was being run as a top-down bureaucracy focused on cutting costs or increasing the current stock price. Such a dynamic undermines and, over time, kills Agile management.

The problem is widespread, even in organizations that are actively embracing Agile at the team level. We found in surveys of Agile teams that some 80 percent to 90 percent of Agile teams perceive tension between the way the Agile team is run and the way the whole organization is run. In half of those cases, the tension was "serious."

The Law of the Network is the current frontier of the Agile movement—how to make the whole organization Agile. It's a tough nut to crack because Agile represents a radically different concept of an organization. At the heart of twentieth-century management thinking is the notion of a corporation as an efficient steady-state machine aimed at exploiting its existing business model. As Google executives Eric Schmidt and Jonathan Rosenberg write in their book, *How Google Works*, "Traditional, MBA-style thinking dictates that you build up a sustainable competitive advantage over rivals and then close the fortress and defend it with boiling oil and flaming arrows."[14]

The fortress is run from the top, with an assumption that the top knows best. The fortress is "built to minimize risk and keep people in their boxes and silos," as business school professor John Kotter writes. People "are working with a system that is designed to get today's job done—a system that asks most people, usually benignly, to be quiet, take orders, and do their jobs in a repetitive way."[15] *Exploitation* of the existing business model takes precedence over the *exploration* of new possibilities.

Over many decades, multiple fixes were explored to alleviate the static nature of the organization, including task forces, special project groups, strategy departments, tiger teams, skunk works, R&D, dual operating systems, knowledge funnels, design thinking, and so on.[16] But these were still fixes to the same concept of the corporation as a static machine with a vertical reporting dynamic. Big bosses continued to appoint little bosses, and so on down the line. The organization continued to operate like a giant warship—big and disciplined but slow and hard to maneuver.

By contrast, when the whole organization truly embraces Agile, the organization is less like a giant warship and more like a flotilla of tiny speedboats. Instead of a steady-state machine, the organization is an organic living network of high-performance teams. In these organizations, managers recognize that competence resides throughout the organization and that innovation can come from anywhere. The whole organization, including the top, is obsessed with delivering more value to customers. Agile teams take initiative on their own and interact with other Agile teams to solve common problems. In effect, the whole organization shares a common mindset in which the organization is viewed and operated as a network of high-performance teams.

■

One common misunderstanding is that Agile organizations are necessarily flat or nonhierarchical. In Agile organizations, the top management still has the important function of setting direction for the organization. People are still removed for not getting their job done. If anything, the drive for higher performance in an Agile organization is even more relentless than in a bureaucracy. Because of radical transparency and peer-to-peer accountability, there is nowhere to hide. Everyone knows everything.

But the hierarchy in an Agile organization is primarily a hierarchy of competence, not a hierarchy of authority. The performance question is not whether you have pleased your manager by doing what you were told to do: The question is whether you have added value to your real boss—the customer. The organization operates with an interactive communication dynamic, both horizontally and vertically. Anyone can talk to anyone. Ideas can come from anywhere, including customers. As a network, the organization becomes a growing, learning, adapting living organism that is in constant flux to identify and develop new opportunities that add new value for customers. When done right, continuous delivery of more value to customers from less work results in generous returns to the organization that provides it.

Agile thus explodes the distinction between exploitation and exploration. All parts of the organization are continuously exploring how to add more value to customers.

In the early years of the Agile movement, critics said that small teams would never be able to handle big, complex problems. It turns out that once the teams are housed in a network driven by horizontal conversations focused on a common goal and operating in a common cadence, then networks of small teams can handle large, complex problems with the same agility as small teams—and much better than a top-down bureaucracy.

■

These three laws—first, small teams working on small tasks in short iterative work cycles delivering value to customers; second, an obsession with continuously adding more value for customers; and third, coordinating work in an interactive network—are the same three principles that enable Spotify to provide personalized music playlists to over 100 million users every week. These same principles are allowing Barclays to become a bank that can provide easy, quick, convenient, personalized banking at scale; and Ericsson's network management to deliver more value to its clients sooner.

When these three laws are in effect, people in the organization share a different way of understanding how the world works and how to interact with the world to get things done. Counterintuitive ideas abound: Managers can't tell people what to do; firms make more money by not focusing on making money; dealing with big issues requires small teams, small tasks, short work cycles—in effect, small everything; control is enhanced by letting go of control.

When a traditional manager encounters an organization with these strange ideas, it's like visiting a foreign country where everything is different, where yes may mean no, where fixed prices are negotiable, and where laughter may mean anger.[17] The familiar cues that enable travelers to function in their home country are absent. In their place are new rules that are odd and incomprehensible. The result can be bewilderment, frustration, and an inability to cope. Until the travelers grasp what has happened, learn the new cues of the different country, and embody them in their behavior, they will feel disoriented and incompetent.

This is why Agile cannot be implemented within the assumptions of current management practice. Agile means embracing fundamentally different assumptions. For traditional managers, the process usually isn't comfortable. It isn't easy. At the outset, it feels wrong. It's like learning a foreign language with a different grammar. It is only over time and through actual experience and practice that Agile becomes second nature and automatic. This is not about "implementing Agile tools and processes." It's about more fully embodying the Agile mindset and acquiring the associated muscle memory.

Ultimately, Agile is about understanding and interacting with the world with a different mindset. When people don't have an Agile mindset, our research shows that even if they are implementing every tool and process and practice exactly according to the book, no benefits flow. Conversely, when people in the organization have an Agile mindset, it hardly matters exactly what tools, processes, and practices they are using. The mindset makes things come out right. In the end, Agile is a mindset.

■

Of the three laws, the first one—the notion that work in principle should be done in small teams working in short cycles—is the best known in the Agile world because that's what received most of the attention of the early Agile implementations.

But it's the second law—the idea that the very purpose of a firm is to deliver value to the customer—that is the most important, because this is the principle that makes sense of the other two and that permits the greatest insight into why an Agile organization operates the way it does.

Yet, the lynchpin of Agile management is really the third law: The impact of high-performance teams with a customer focus is suboptimal unless and until the whole organization operates as an interactive network. It is when the three laws combine together and focus on a common external goal that we get the explosive increment in value that comes from truly embracing Agile management.

Agile management thus operates under three laws, and together they generate the basics of the Agile organization. The three laws allow us to make sense of the myriad Agile practices that may—or may not—be applicable in any particular context. Practices may change, but the Agile mindset applying the three laws endures. The Agile mindset offers a lasting guide to what's involved in an organization implementing the new management paradigm.

BOX 1-1

MANIFESTO FOR
AGILE SOFTWARE DEVELOPMENT

We are uncovering better ways of developing software by doing it and helping others do it. Through this work, we have come to value:

- ♦ Individuals and interactions over processes and tools
- ♦ Working software over comprehensive documentation
- ♦ Customer collaboration over contract negotiation
- ♦ Responding to change over following a plan

That is, while there is value in the items on the right, we value the items on the left more.

We follow these principles:

1. Our highest priority is to satisfy the customer through early and continuous delivery of valuable software.
2. Welcome changing requirements, even late in development. Agile processes harness change for the customer's competitive advantage.
3. Deliver working software frequently, from a couple of weeks to a couple of months, with a preference to the shorter timescale.

4. Business people and developers must work together daily throughout the project.
5. Build projects around motivated individuals. Give them the environment and support they need, and trust them to get the job done.
6. The most efficient and effective method of conveying information to and within a development team is face-to-face conversation.
7. Working software is the primary measure of progress.
8. Agile processes promote sustainable development. The sponsors, developers, and users should be able to maintain a constant pace indefinitely.
9. Continuous attention to technical excellence and good design enhances agility.
10. Simplicity—the art of maximizing the amount of work done—is essential.
11. The best architectures, requirements, and designs emerge from self-organizing teams.
12. At regular intervals, the team reflects on how to become more effective, then tunes and adjusts its behavior accordingly.

Kent Beck	James Grenning	Robert C. Martin
Mike Beedle	Jim Highsmith	Steve Mellor
Arie van Bennekum	Andrew Hunt	Ken Schwaber
Alistair Cockburn	Ron Jeffries	Jeff Sutherland
Ward Cunningham	Jon Kern	Dave Thomas
Martin Fowler	Brian Marick	

BOX 1-2

GLOSSARY: DEFINITIONS OF AGILE, SCRUM, DEVOPS, KANBAN, LEAN

Agile is a movement that took off in 2001 as a set of values and principles articulated by the Agile Manifesto of 2001, although it had many earlier antecedents, such as "quality" and "design thinking." In due course, the Manifesto spawned various management methodologies including Scrum, DevOps, Lean, and Kanban. Over time, it evolved into a movement of people with a specific mindset. The mindset focuses on delivering continuous value to customers as the primary goal of work. It embraces iterative, incremental approaches to working in small teams and aims at enterprise-wide agility by operating as a network.

Scrum is the principal Agile management methodology. It uses a cross-functional team-based approach for delivering value to organizations and customers, with specific roles for the Product Owner and the Scrum Master. The team respects individual contribution and builds on the strengths of accountability, deep interpersonal relationships, collaboration, and teamwork. Managers are no longer bosses, but coaches who remove impediments and clear the way for teams to provide value to their customers by remaining focused and creative.

DevOps (a clipped compound word that combines development and operations) is a culture, movement, and practice that emphasizes the collaboration and communication of both software developers and other information technology professionals while automating the process of software delivery and infrastructure changes, with very rapid deployment of changes.

Kanban is a scheduling system for software development, lean manufacturing, and just-in-time manufacturing. Kanban can also serve as an inventory-control system to control the supply chain. One of the main benefits of kanban is that it establishes an upper limit to the work in process, avoiding overloading of the system.

Lean is a systematic methodology for the elimination of waste within a system of manufacturing or software. Essentially, lean is centered on making obvious what adds value by reducing everything else.

Lean Startup is a methodology for developing businesses and products based on the hypothesis that if firms invest their time into iteratively exploring the needs of early customers, they can reduce market risk, reduce the need for initial project funding, and enhance the chance of ultimate success.

Design Thinking is a human-centered approach to innovation that seeks to integrate the needs of people, the possibilities of technology, and the requirements for business success. Design thinking has now spread beyond professional design practice and is applied to a wide range of businesses and to social issues.[1]

NOTE

1. Sources include Herbert A. Simon, *The Sciences of the Artificial* (1969); R. McKim, *Experiences in Visual Thinking* (1973); B. Lawson, *How Designers Think* (1980); and R. Buchanan, "Wicked Problems in Design Thinking," *Design Issues* 8, no. 2 (Spring 1992). Design thinking was further adapted for business purposes by the design consultancy IDEO.

2.

THE LAW OF THE SMALL TEAM

The fundamental task is to achieve smallness within large organizations.

—E. F. SCHUMACHER[1]

icture this. It's 1997 and you've just had a great idea: a slim hand-held device that's small enough to slip into your pocket and that can perform multiple functions at the touch of a fingertip. It will serve as a portable telephone, an address book, a map, a navigation system, an airline boarding pass, a music or movie player, a television set, a camera, a flashlight, a dictation machine, a stopwatch, an alarm clock, a translation machine, a remote control, a repository of the world's newspapers and magazines, a library of thousands of books, and more. Moreover, the device will be lightning-quick in its responsiveness and it will be customizable to meet the individual needs and preferences of hundreds of millions of people around the world.

Sound good? You know it's a winner. But how to bring it to reality? Well, it's 1997 and so you apply the best management practices of the day. First, you spend a couple of years winning support for the idea's feasibility and persuading the strategy committee of a major corporation to design, build, and market this amazing new device. After finally getting this approval, you pull together a huge team of designers and engineers and spend another couple of years developing detailed specifications for the device with a timeline to deliver and integrate its various components. You then recruit hundreds of thousands of engineers and developers to build the device. You also hire tens of thousands of managers to supervise and control them to ensure that they deliver according to the plan and on schedule. You set up comprehensive reporting systems so that top management knows exactly how every dollar has been spent and what progress is being made on each component, thereby eliminating risk. You put in place a system of coordination committees to ensure that all the many units will be working together on the device like a finely tuned symphony orchestra.

The result? Many years and billions of dollars later, you find that your device is still far from being ready for the marketplace. The technical problems are unending and seemingly exponential. Coordination between units is a massive headache, despite vast amounts of time and effort by the coordination committees. There is acrimonious finger-pointing about who is responsible for the technical problems and the delays. While many individual components appear promising, they are often incompatible with each other. Each unit blames the other, but in the end, no one can be found accountable and fixing the problems takes forever. Even when solutions to known problems are produced, they generate fresh problems. Worse, you suspect that there's a massive backlog of unknown technical problems waiting to explode at any moment. And your device is still years away from being ready for the marketplace.

In organizing the work this way, you had hoped to achieve economies of scale. But you see now that you have achieved the opposite: *dis-economies of scale*. The cost associated with organizing and coordinating so many people and units is growing more quickly than any additional value that those people are likely to create. And whether they will ever complete the work is still a question mark.[2]

Sadly, after years of effort and billions of dollars spent, management decides that your project must be canceled. That's because a real-life competitor—Apple—had by 2007 developed a similar product, having brought it "from scratch to market" in just eighteen months at a fraction of the cost of traditional management practices.

How was that possible? Instead of building a very complex device with a very complex organization according to elaborate specifications, Apple did the opposite. It designed and built a relatively simple physical device—the iPhone—that was developed in an iterative fashion in short cycles and steadily upgraded. Instead of recruiting hundreds of thousands of its own engineers to build a monolithic software operating system to perform the device's functions, Apple set up a technology platform and invited hundreds of thousands of independent teams of developers to imagine and create their own purpose-built applications ("apps") to meet every conceivable human need and to offer them directly to Apple's customers. Together, the apps can perform an almost infinite array of functions. They are added as they are developed while interacting directly with customers. The customers themselves decide which apps are valuable for their own specific needs and configure the device according to their own idiosyncratic preferences. The result is a multifunctional device that is customized to meet the needs and passing whims of hundreds of millions of individual users—a feat inconceivable with traditional management methods. Instead of *scaling up* the organization to resolve a complex problem, Apple *scaled down* the problem into tiny bite-size pieces that small independent teams could deliver in an iterative fashion with direct feedback from customers.

The success of Apple's iPhone is often attributed to the genius of Steve Jobs. Brilliant marketing. Superior design. Meticulous attention to detail. Breakthrough thinking. Ferocious drive to solve problems. All of this is true. But what is often overlooked is that these elements would have gone for naught if Apple had not developed a relatively simple hardware device in an iterative fashion and offered a platform to mobilize independent software developers: hundreds of thousands of small teams of developers contributing their ingenuity and talents to build apps in an iterative fashion while interacting directly with customers.

■

The Law of the Small Team is simple. It's a presumption that in a VUCA world, big and difficult problems should—to the extent possible—be disaggregated into small batches and performed by small cross-functional autonomous teams working iteratively in short cycles in a state of flow, with fast feedback from customers and end-users. Instead of constructing a big and complex organization to handle complexity, the organization disaggregates the problem into tiny pieces so that it can be put together in minuscule increments and adjusted in the light of new and rapidly changing information about both the technology and the customer.

The lessons of the Law of the Small Team are still being learned. The failed effort to produce a portable device by traditional management practices (as described in the opening story) is imaginary, although some aspects are eerily reminiscent of the history of the Newton—a personal digital assistant that Apple's then CEO, John Sculley, oversaw development of in 1987 until it was abandoned by Steve Jobs in 1998.[3]

The real-world disasters that flow from trying to manage complexity within a top-down bureaucracy are very real. For instance, in 2006, the U.S. Air Force launched a project aimed at modernizing the management of logistics. It awarded a $628 million contract to Computer Sciences Corporation to serve as the system integrator; its job was to configure, deploy, and conduct training and change management activities before the launch.[4]

"We've never tried to change all the processes, tools and languages of all 250,000 people in our business at once," said Grover Dunn, the Air Force director of transformation, "and that's essentially what we're about to do."[5] As Elizabeth McGrath, the Defense Department's deputy chief management officer, explains, "We started with a Big Bang approach and put every possible requirement into the program, which made it very large and very complex."[6]

Over the next seven years, the project was restructured several times. Finally, in 2013, the Air Force realized that it would cost another $1 billion to achieve one-quarter of the capabilities originally planned. Even then the system would not be ready for another seven years. So, the Air Force gave up on the project entirely. It canceled the project after spending some $1.3 billion.

■

You might say that the approach of disaggregating work into tiny pieces might make sense for a personal entertainment device like the iPhone. But could it ever really work for a serious industrial project where reliability is required—such as, say, a stealth fighter jet? The answer is that it already has. The Swedish aircraft maker Saab has done exactly that. Saab's Gripen fighter jet was developed and built using Agile practices.[7] Every six months Saab issues a new release of the jet's operating system that makes it faster, cheaper, lighter, more efficient, more powerful, with better electronics and more sophisticated targeting. Serious defense experts have called it "the best stealth fighter in the world."[8]

A study conducted by the respected international defense publishing group IHS Jane's compared the operating costs of Saab's Gripen fighter aircraft with Lockheed Martin's F-16 and F-35 aircraft, Boeing's F/A-18 Super Hornet, Dassault's Rafale, and Eurofighter's Typhoon. It concluded that the Gripen had "the lowest operational cost of all these aircraft in terms of fuel used, pre-flight preparation and repair, and scheduled airfield-level maintenance together with associated personnel costs."[9]

Software plays a steadily increasing role in the design and evolution of the Gripen: "The conundrum facing fighter planners," writes Bill Sweetman in *Aviation Week and Space Technology*, "is that, however smart your engineering, these aircraft are expensive to design and build, and have a cradle-to-grave product life that is far beyond either the political or technological horizon." Saab's Gripen has been designed with these issues in mind. "Long life requires adaptability, both across missions and through-life."[10]

The same phenomenon is emerging in the world of automobiles. For the first century of the history of cars, when you bought a car, you were stuck with what you bought. If you wanted something better, like more features or more power, you had to buy a new car. Now that's changing. Tesla, for instance, can add new capabilities into cars it has already sold by downloading new software for the car. The new features include automatic braking when the vehicle senses a pending collision, a partial autopilot system, and a robotic parking program. These features aren't unique to Tesla's cars—they are found on other luxury cars like the Audi or Mercedes-Benz. What's different is that the Tesla Model S is designed to be continuously upgraded on the fly. Tesla can remotely add substantial functions to cars already on the road.

"We really designed the Model S to be a very sophisticated computer on wheels," says CEO Elon Musk. "Tesla is a software company as much as it is a hardware company. A huge part of what Tesla is, is a Silicon Valley software company. We view this the same as updating your phone or your laptop."[11]

Cars are thus more and more resembling flexible electronic devices than mechanical machines. Like the iPhone, the car is becoming a platform for apps that steadily improve the vehicle's functions rather than embodying static performance defined by mechanical features installed at the time of purchase.

∎

In understanding the scope of Agile, it's useful to remember that Agile management became associated with software only after 2001, and that its historical roots lie in manufacturing with the quality movement in Japan and the Toyota Production System. Toyota began by disaggregating the manufacturing process itself. It experimented with small production runs. Contrary to what common sense might indicate, Toyota discovered that once rapid changeovers were accomplished, small demand-driven iterations generally turned out to be more efficient than mass-production runs.[12]

This model of manufacturing spread throughout Japan in the 1970s and came to the United States in the 1980s. It was discovered that when these iterative techniques were executed well, with small batch sizes, cycle times could drop by factors of 10 to 100. Inventories could be reduced by more than 90 percent, freeing enormous amounts of cash. Secondary effects include improved quality, accelerated learning, and lower production costs.

Iterative thinking was then applied to the process of developing new products, as described in the seminal 1986 article in *Harvard Business Review* by Hirotaka Takeuchi and Ikujiro Nonaka, "The New New Product Development Game." The authors wrote:

> Companies are increasingly realizing that the old, sequential approach to developing new products simply won't get the job done. Instead, companies in Japan and the United States are using a holistic method—as in rugby, the ball gets passed within the team as it moves as a unit up the field.
>
> This holistic approach has six characteristics: built-in instability, self-organizing project teams, overlapping development phases, "multi-learning," subtle control, and organizational transfer of learning. The six pieces fit together like a jigsaw puzzle, forming a fast, flexible process for new product development. Just as important, the new approach can act as a change agent: it is a vehicle for introducing creative, market-driven ideas and processes into an old, rigid organization.[13]

The examples given in the article—Fuji-Xerox, Honda, and Canon—are all hardware, not software.

In 1990, the iterative small-team approach was further disseminated as "lean manufacturing" in the classic book *The Machine That Changed the World*.[14] But while the systematic use of small teams and iterative approaches began in hardware, it really took off in software development after the publication of the Agile Manifesto in 2001.

■

What *exactly* are the practices that make up the Law of the Small Team? One answer is that "it depends." For the first decade after the Agile Manifesto, there were many furious arguments among Agile practitioners as to what are the "true Agile practices." Some urged Scrum. Others were adamant that kanban was the answer. Still others were convinced that the answer lay in lean manufacturing. Eventually it became clear that the answer to the question was something different altogether. The Law of the Small Team concerns a *mindset*, not a specific set of tools and processes that can be written down in an operational manual. If you are thinking about Agile as a set of tools and processes, you're looking for the wrong thing. You can't go to the store and "buy some Agile management."

The Law of the Small Team is a *presumption* about how complex work in principle should be done. In any particular organization, the practices that emerge will be the result of an interaction between the Agile mindset and the specific organizational context. That's one reason why hiring a firm of consultants to "come in and train our staff on Agile management tools and processes" rarely by itself produces a happy result.

Over the last couple of years, the SD Learning Consortium has organized a series of mutual site visits among its members to learn what is involved in successfully implementing Agile management.[15] The object is for the firms to find out from each other what Agile implementation looks like in real life. The result? In each successful case, we found that the firm started from some general principles and prior examples and then organically grew a set of practices to meet its own specific needs and culture, sometimes with its own idiosyncratic labels. Although there is no "one size fits all" and no universal "best practice," we did see convergence toward management practices that have a striking family resemblance. The main characteristics are listed in Figure 2-1.

[FIGURE 2-1]

COMMON PRACTICES OF
AGILE SMALL TEAMS

1. *Work in small batches.* To cope with complexity and unpredictability, work is (to the extent possible) broken down and disaggregated into batches in which something potentially of value to a customer or end-user can be completed in a short cycle. By having teams working in short cycles, it is easy to see, even in large complex projects, whether progress is being made—or not. In some cases, the firm prescribes a common cadence, usually one, two, or three weeks, while in other cases each team is free to select the appropriate periodicity.

   These firms had all seen big and complex plans fail because there were too many unknowns and change was happening too quickly for adjustments to be made. The response has been to think differently: small batches of work, small teams, short cycles, and quick feedback—in effect, "small everything."

2. *Small cross-functional teams.* Work is typically done by small, autonomous, cross-functional teams that can complete something potentially of value to a customer. The size of the teams varies. One rule of thumb is "seven plus or minus two." In some firms, the teams have ten to twelve people. In other cases, the teams are smaller. Sometimes the teams have different names, like "pods" or "squads," and the word "team" is applied to the larger project that the small groups are working on.

3. *Limited work in process.* In Agile management, teams learn to focus on an amount of work that can be brought to completion in each short cycle. By limiting the amount of work in process at any one time, the risk of work waiting in queues is reduced. Excessive work-in-progress is a pervasive feature in teams getting started in Agile and in back-office functions where work tends to accumulate in queues.

4. *Autonomous teams.* Once it is decided at the beginning of each short cycle *what* to do, teams themselves decide *how* to get work done. In each case, the firm decides some basic "rules of the road," but after that, the team has autonomy on how to proceed. The "rules of the road" vary from firm to firm. Some firms implement arrangements akin to Scrum, in sprints, with a common cadence to enhance the capability of managing dependencies between teams. In other firms, those choices are left to the team. In all firms, we saw provisions as to how the team is led and the accountabilities of the team. But *how* the work is actually done is, in each case, up to the team.

5. *Getting to "done."* A common litmus test of successful Agile implementations is whether the teams are routinely producing fully finished work at the end of each cycle. Keeping batch sizes small helps teams get work fully "done," not just "almost finished." The idea of getting to "done" sounds absurdly simple, but it turns out to be transformative. One reason big bureaucracies are so slow is that that they have vast amounts of partly finished tasks, often with hidden unresolved problems, all of which creates additional work when tasks are resumed: Context switching is an expensive cognitive function.

   In software development, a common definition of "done" includes completed code, unit tests completed, integration tests completed, and performance tested and approved by the customer. This is very difficult to accomplish if the team is working on a large task. By keeping the task small, transparency is facilitated. By achieving problem-free work at the end of each short work cycle that can potentially be tested on a customer, snags and snafus are identified early and technical debt doesn't accumulate.

6. *Work without interruption.* Within each short cycle, teams pursue their work without interruption. Once it is established at the start of the short cycle what is high priority, the presumption is that managers and the team stick with that decision for the duration of the cycle.

(continued on next page)

(continued from previous page)

7. *Daily standups.* Daily standups were observed as a universal ritual of all the site visits, whatever the particular Agile practices in use. In the daily standup, teams hold brief daily meetings to share progress and identify impediments for removal. The topics vary somewhat but typically concern what work has been done, what will be done next, and what impediments are being experienced. Standups help the members of the team "swarm" to solve a problem rather than have individuals struggle on their own. The communications are intended for the team members themselves, not for managers to inspect and control the progress of the team.

8. *Radical transparency.* The use of "paper-based information radiators" was striking during all the site visits. In effect, anyone can walk into a team space and see at a glance what the status of the work is and where any problems may lie.

9. *Customer feedback each cycle.* Teams receive feedback from the customer at the end of each short cycle. In collaboration with managers, teams evaluate what has been accomplished in the light of feedback from customers and incorporate the feedback into the planning of next steps.

10. *Retrospective reviews.* Retrospective reviews of what has been learned occur at the end of each short cycle and provide a basis for planning the next cycle of work. As in the daily standups, the conversation is intended for the team members themselves, not for managers to inspect and control the progress of the team.

This way of working emerged out of practical experience and extensive experimentation. Is it more productive? Google thinks so. At Google, "the company's top executives long believed that building the best teams meant combining the best people," said Abeer Dubey, a manager in Google's People Analytics division. But when they studied it, it turned out not to be the case. A comprehensive study of what characteristics led to a high-performance team showed that there was practically no correlation with the types of individuals on the team. The study, which became known as the Aristotle project, showed that the composition of the team had much less to do with team performance than five key dynamics that are supported by Agile management practices.[16]

1. *Psychological safety.* Can we take risks on this team without feeling insecure or embarrassed?
2. *Dependability.* Can we count on each other to do high-quality work on time?
3. *Structure and clarity.* Are goals, roles, and execution plans on our team clear?
4. *Meaning of work.* Are we working on something that is personally important for each of us?
5. *Impact of work.* Do we fundamentally believe that the work we're doing matters?

The result of organizing work in an Agile manner is not only that work gets done faster, but that those doing the work tend to be engaged in what they are doing. Unlike the traditional workplace where only one in five employees is fully engaged in his or her work, in an Agile management workplace, employees are in a psychological state of "flow," as identified by Mihály Csíkszentmihályi—that is, those doing the work are fully immersed in a feeling of energized focus, full involvement, and enjoyment in the process of the activity. When those doing the work can see the meaning of what they do for those for whom they are doing it, they "bring their brain to work."

This phenomenon accounts for the high motivation that we observed in teams encountered during the site visits. When we asked staff members in these organizations whether they would ever willingly work in a firm using traditional management practices, the universal answer was "never."

■

Although there are strong family resemblances among the practices, each firm has developed its own unique combination of practices. An interesting mix occurs at Menlo Innovations, a software design and development firm in Ann Arbor, Michigan, and the brainchild of its hugely enthusiastic CEO, Richard Sheridan, and cofounder James Goebel. Menlo delivers custom software applications, mostly for business-to-business purposes and often in mission-critical fields such as medical and health care. What's nice is that Menlo welcomes visitors. You can go and visit the firm in Ann Arbor and observe this workplace of the future.

Sheridan talks more about joy in the workplace than about Agile. So, when visitors to Menlo come to learn about Agile and lean, why does he talk about joy? "It's simple," says Sheridan. "What we have tried to do at Menlo is to emancipate the heart of the engineer, which is to serve others. We engineers exist to produce something that the world will enjoy, something that will delight people. All of these things that we call Agile or lean, or any of the other processes that have a name, that's what they are really about when they are done well: How do we serve others?

"In doing so, we borrowed from a lot of concepts," Sheridan says. "That's why you won't catch us using the word 'Agile' very often. People visit us and say, 'You look very Agile to us. Why not call it Agile?' We are not doing these things because we want to be Agile. We're doing them because we want to produce joy in the world with our work."

It all began some twenty years ago when Sheridan was exasperated with the bad organizational practices he had experienced as a manager in software development. In his own firm, he set out with the explicit goal of ending the human suffering involved in software development.[17]

Before he set up Menlo Innovations, almost the *only* thing he did was to resolve software crises when systems broke down or experienced major outages. Traditional management processes were creating fires and Sheridan had to put them out. He felt like a chief firefighter. The stress level was terrible. So, he set out to create a firm that didn't do that to its people—a firm that wouldn't experience constant emergencies with high levels of angst. He wanted people to be proud of what they were doing and how they were doing it. He wanted software to be produced by people who experienced joy in what they did, not stress.

The word "joy" is not one normally associated with the twentieth-century workplace. Good workplaces were perhaps "effective" or "efficient" or even "engaged." But joyful? Even thinking about the possibility initially sounded ridiculous.

Yet for some fifteen years, Menlo Innovations has been doing exactly that—maintaining a workplace that generates joy both for those doing the work (software developers), for those for whom the work is being done (clients), and for those who will one day use the results of the labor (the end-users). "We don't believe for a second," says Sheridan, "that we can serve others well unless we are taking care of our team. So, we created a system that is focused first and foremost on servicing others and producing great results for the world. But in order to do that, we had to recraft the way software is designed and developed."

Sheridan's goal of producing a joyful workplace means generating ultra-reliable code that never breaks and never needs fixing. This in turn reflects the sector in which Menlo operates—medical devices that *must* be fail-safe. "The last time we had a software emergency was in 2004," says Sheridan. "It's different with some of our clients. We just started work with a new client and in the very first week, the internal team launched a big new project that had been under development for years and it was a disaster. People were working nights and weekends, trying to recover. It affected their business in big ways and slowed sales of their products in the marketplace."

By contrast, Menlo operates as an emergency-free environment. When Menlo's staff saw what was happening with this client, they said, "Oh, so that's what an emergency looks like!" They had never seen one. They had never experienced it. They didn't know what it was.

"What we are doing," says Sheridan, "is creating freedom through tyranny. It's true that we have created a joyful place and people love working here and they are fully engaged. And yet it's also true that we have introduced tyranny by removing ambiguity from the workplace. In our world, people know who they are working with and who they are working for. They know what they are working on, and they know what order they are going to work on it. That's the tyranny part. Once that's established, the freedom part kicks in. I say, 'You are now free to pursue the work that you love without anyone hanging over your shoulder, cutting in, and asking what you are working on and how it's going.'

"We have a process," says Sheridan, "that the team believes in. And because they believe in the process, we don't need process police going around checking whether they are staying on task and doing their jobs. The process they know actually protects them. The process is clear both to them and to our customers. Teamwork is not optional. Collaboration is not optional. We say that we are giving our teams 'permission to collaborate.' By forcing the idea of working in small teams, we are creating the freedom to collaborate."

How does Menlo produce "emergency-free software"? One striking feature is how Menlo approaches the issue of team continuity. Most firms practicing Agile management, such as Microsoft and Ericsson, put effort into keeping the members of the teams together, with the same people working on the same team as long as possible. The idea is that such teams become like a champion sports team in which everyone knows each other's skills and idiosyncrasies. Moreover, the ones who create the code are the ones who understand it and who are best placed to maintain it. Giving them responsibility for fixing their own quality problems ("bugs") also gives them an incentive to get it right the first time. These firms almost see the team itself as the product.

"That works fine," says Sheridan with a smile, "as long as the team is immortal and people never take vacations! We do the opposite because we know that if software is dependent on the individual members of a team, we have to ask: What happens if that team [member] is not available? If there are four people working on a project in a traditional setting, each of them will be a tower of exclusive knowledge and expertise in their own area. As a result, you can't operate in the absence of one of the experts without the entire team crashing. You have to have the database guy. You can't lose the secret sauce guy. You can't move without the middleware guy. That's because they are the only ones who know their own piece of the code. In effect, the knowledge about that piece of software is trapped inside the team. We need teams and software that are not dependent on individuals."

At Menlo, the work cycles are very short—only one week—and the membership of teams is deliberately changed each week. "Because of our switching of team members, each team is touching code that someone else wrote the previous week. Effectively, we are testing right then and there whether the code is reliable and maintainable. The new team needs to understand the code in order to make changes effectively. If they don't understand the code, they can go and find the pair that wrote the code. They might be sitting just a few feet away. In the process, the first team also gets to understand the code that they created and discover what they thought was clear wasn't clear after all.

"The result," says Sheridan, "is really clear, reliable, maintainable code. That's our goal. When we work this way over and over again every single week, we end up with a rock-solid body of code that is clear and maintainable by lots of people. That enables us to expand and extend the code. It gives us code that is emergency-free."

A second unusual feature at Menlo is the approach to staffing. Many firms approach staffing by trying to find exactly the right person with exactly the skills they need, such as Python 4.6 or Oracle 9.1.1.1 or whatever. To solve this problem, they try to match the skills needed for a particular task with a person who possesses those precise skills.

Menlo does the opposite. It seeks to create a culture of collaboration, adaptability, transparency, and trust. "It starts with our hiring practices," says Sheridan. "Unless our people fit the culture, we won't have a chance of maintaining our culture over time. When we need new staff, we bring people in and get them to 'speed date' with our own staff. The question is always: Would you like to work with this person? If the answer is yes, then we bring them in to work with us for a day, then a week, and then a month. If the answer is still 'Yes, I would like to work with this person,' then they are hired."

Says Sheridan, "We developed this interview process to match our culture. In the beginning, we don't ask many questions. We don't pay too much attention to résumés. We are more interested in culture fit. We don't start with how smart you are. We don't care what you've learned. We aren't interested in where you've gone to college. None of that matters if you don't fit our culture. Once we've determined that there is a fit with our culture, then we start to ask about skills. But frankly, with the right culture, acquiring skills is not all that difficult in comparison to having the right attitude and mindset."

A third striking feature at Menlo is the use of pairing: All work is done in pairs. To traditional managers, having two people working on the same computer and doing the work of one is unproductive and inefficient, almost by definition. Managers think pairing must be cutting productivity in half. Sheridan says it's the opposite. Menlo does all its work in pairs precisely because it's so productive. Performance measures show that it is up to ten times more productive than working as individuals.

Earlier in his career, Sheridan would have bosses who, when they saw two people working together, would assume that at least one of them wasn't working at that moment, and probably both of them. They would inject themselves into the conversation and make those people "go back to work." Menlo is the opposite of that. "It's only if we see someone alone," says Sheridan, "that we are likely to ask: 'Why aren't you working?'"

Menlo is practicing code *stewardship* of maintainable code, rather than code *ownership* that is hard to maintain. Menlo isn't the only firm practicing pairing. But it's still rare.

■

A fourth striking feature of Menlo is the central role played by the anthropologists. "Sadly, in software we have created an industry," says Sheridan, "which produces technology that doesn't fit people's needs and then treats them as 'stupid users.' We write guides for people showing them how to use software that doesn't meet their needs very well. In Menlo, we take exactly the opposite approach. We want to create software that doesn't require user manuals or help desks. We aim to delight the people who are going to use the software that we are developing. The way we do this is by learning a great deal about the people we serve—those people who are going to touch the software we create. We call this practice 'high-tech anthropology.'

"Our anthropologists," says Sheridan, "using empathy and compassion and the tools of the anthropologists, go out into the world. They go to where the work is done and not only understand the people and the environment, but also the vocabulary, the workflow, the habits, the fears, and the dreams of those human beings, and [they] design a software experience accordingly."

These four atypical practices at Menlo—recruiting, pairing, changing team membership, and using anthropologists—are not being presented here as "better" than the practices at other organizations or something that every firm should emulate. They happen to fit Menlo's context and culture. They may or may not fit other organizations. They are options that other organizations may—or may not—adopt. One size does not fit all.

The Law of the Small Team involves working in short cycles. Microsoft and Ericsson have adopted three-week cycles. Menlo works in one-week cycles. How short can a cycle be? When I heard about Etsy, the rapidly growing billion-dollar online market for handicraft products that deploys more than *thirty innovations each day*, my immediate reaction was that this must be a hellishly stressful place to work. I was therefore surprised to learn that it is the opposite. Continuous deployment was introduced not only to accelerate innovation but also to overcome excessive stress in the workplace. It has done both.

Seven years ago, Etsy was issuing code changes every two to three weeks, which was already quite rapid compared to what other firms were doing at the time, such as Salesforce's three releases per year or the Microsoft Office upgrades that used to be issued every couple of years. Like many firms, Etsy had separate groups for the development and the deployment of software. The developers created new software and operators would deploy the bundle of changes that the developers had put together. The work environment was characterized by long days and late nights, particularly when something went wrong, which was happening all the time. There were frequent and prolonged outages. Some of this resulted from the fact that software was being written by one group of people and then introduced and operated by a different group.

In 2010, the management at Etsy set out to fix this. It was committed to "innovate or die" and wanted to resolve the hurdles of improving and scaling a website that had become huge, with more than 60 million unique visitors per month. Etsy managers saw that quality involved not only releasing high-quality code, but also improving adaptability and response time when there were problems. They also wanted to have a healthy engineering team working at a sustainable pace. Like many Agile firms, they were committed to Dan Pink's principles of "autonomy, mastery and purpose."[18] Above all, they wanted to reduce the stress levels that were associated with releases of upgrades.

They had put in place continuous integration and automated testing that, in theory, ensured that the effects of any change in the code could be ascertained in a staging area that was an exact replica of the operating site. But in practice, unexpected things kept happening in deployment. Users sometimes did unexpected things, like launching a flash sale that inundated the site with a huge number of users at one instant. Or there could be some unexpected interaction between the actual hardware and the software. Or sometimes in a bundle of many changes it was difficult to figure out exactly what was the cause of the problem.

Etsy decided to go really small. They found—to their surprise—that by having more frequent releases of small changes, it was easier to spot and fix problems. To make the shift happen, basic changes in the implementation and the management approval process were needed. Instead of management approval being required for all changes to the actual site, now all improvements that had been tested were deployed immediately. The staff devising the improvement also oversaw implementation.

Management has articulated a clear vision for the firm and the staff are committed to it, making Etsy a "cool" online market for handmade and vintage items, just as Spotify offers "cool" music streaming. At firms like Etsy and Spotify, purpose isn't a problem. Within the broad strategic framework, the staff are authorized to proceed with continuous improvement. Many of the changes are "dark" changes that users never see, but which improve the performance of the site. There is continuous testing, and if an improvement doesn't make things better for customers, it is immediately removed.

Some changes are experiments aimed at finding out whether something would work better. For instance, would it make sense to include the shipping price in the price of the goods and then offer "free shipping"? The answer to this apparently promising idea in one experiment turned out to be no.

The result is a workplace that is very different from what it was seven years ago. Along with the marked acceleration in innovation, the work is much less stressful. Instead of a separate "deployment army" to cope with bundled changes, now the same staff who are devising the improvement oversee the release, which improves mastery and autonomy. The long days and late nights of the developers at Etsy are largely gone. Also largely gone are prolonged outages. Yes, there's still stress when something goes wrong, but it is much less frequent and much easier to fix. The engineering team now has normal work hours, with evenings and weekends off.

With an average of thirty changes to the website deployed each day, we are looking at extremely rapid innovation. Each of the changes is small, but a small change can be significant, sometimes adding millions of dollars in sales. Doing all this change in tiny increments at warp speed within the framework of a central strategy enables extremely rapid innovation and learning, as well as much greater facility in spotting and fixing any problems that may emerge.

Obviously, this process can only work in an environment with a work culture of trust and collaboration. Is it possible to establish stress-free innovation at warp speed? The experience at Etsy suggests that the answer is yes.

Is it sustainable? In 2015, Etsy went public. Time will tell whether this move has put the organizational culture at risk. Already there are worrying signs of pressure from investors to maximize shareholder value. If Etsy succumbs to those pressures, it may join firms pursuing short-term financial gains at the expense of sustainability. Like Spotify fighting Apple Music, Etsy's only chance of survival is to innovate faster than its competitors and delight its customers to such an extent that Etsy is the must-have experience for steadily more customers.

■

When confronted with Agile teams, traditional managers often scoff. "Teams? Nothing new there. We've always had teams. This is not even remotely a new idea."

In one sense, they are right. As noted in Chapter 1, teams have been talked about in management for almost a century. Yet many of the teams were teams in name only. When the team leader acted like any other boss in a bureaucracy, it was hardly a surprise that few teams achieved consistently high performance.

Teams were generally used for solving specific problems, such as R&D or special projects, but not for day-to-day management. The thinking was that you could have disciplined execution through bureaucracy, or innovation with teams, but not both. You had to choose. (See Figure 2-2.)

The result of the innovations made by Agile management is that there is no longer a choice to be made between disciplined execution and innovation through teams. The new way of operating enables the firm to do both at the same time. Now teams can be deployed to handle almost all work. (See Figure 2-3.)

The twentieth-century humanist managers promoting teams meant well. These psychologists, sociologists, and management consultants

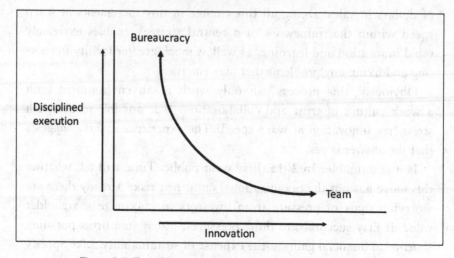

Figure 2-2. Traditional management had to choose disciplined execution vs. innovation.

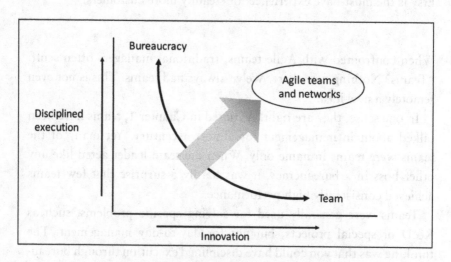

Figure 2-3. Agile management achieves both disciplined execution and innovation.

had analyzed what people needed and how they should work in groups. Surely employees who saw work as meaningful and felt fulfilled in a workplace with a friendly atmosphere would be more productive. Surely managers who were perceived as kind and caring, and who could inspire and coach employees, would do better than bosses who were mean and surly.

The language of teams became pervasive, even if the practices of high-performance teams didn't. There was talk of "team spirit," "winning," self-actualization," "excellence," "commitment to a goal," "the desire for perfection," and so on.

But often the language of teams was a desolate echo of the real thing. It consisted of empty clichés, a rah-rah world of fake conviviality. The actual implementation was thus a drab caricature of the romantic language in which these schemes were couched. The language embodied a fairy tale about the enchanted corporation.[19]

By contrast, the founders of Agile management chose terms that were deliberately mundane and even ugly: "Scrum," "Scrum Masters," "product owners," "kanban," "burndown charts," "working software," "sprints," "standups," and getting work not just "done" but "done, done, done." There was no hint here of magic or enchantment, no highfalutin language about self-actualization and no fake camaraderie. Instead there was a rigorous focus on solving problems and enabling people to get on with work without interruption, drawing on real expertise, removing impediments, and continuously delivering value to customers in the face of mind-boggling complexity.[20]

The result is that Agile teams operated very differently from the team initiatives of the twentieth century.

- Whereas the language of the twentieth-century team initiatives was romantic, the language of Agile teams is down-to-earth and pragmatic.
- Whereas the twentieth-century team initiatives tended to assume linear problems that could be definitively solved, Agile teams accept unfathomable complexity and the nonlinearity of a continually morphing environment as the basic nature of the game.
- Whereas the twentieth-century teams had difficulty coping with large-scale problems, the use of platforms and networks has enabled Agile management to achieve almost infinitely large scale without sclerosis.

These considerations help us see why the Law of the Small Team is more than a change in terminology. By adopting the small team as the default method for getting *almost anything* done, Agile management changed the game fundamentally.

In retrospect, we can see why the problem that the humanist theorists were trying to solve—a zero-sum power struggle between bosses and workers—was insoluble. It was a battle that bosses would inevitably win, although the victory would turn to dust in their hands, since the workers became dispirited.

The voice of the customer was often lost in the struggles up and down the hierarchy, between the competing organizational silos and between differing managerial agendas. What was needed was to change the game from a zero-sum vertical dynamic of conflict into a win-win horizontal dynamic of delivering value to the customer through collaboration and joint problem solving. Whereas the twentieth-century team initiatives were often internally focused, what was needed was a shift to an external focus by providing value continuously for the customer. How is this accomplished in practice? It is to this set of issues that we now turn.

3.

THE LAW OF THE CUSTOMER

All truth passes through three stages. First, it is ridiculed. Second, it is violently opposed. Third, it is accepted as being self-evident.

—ARTHUR SCHOPENHAUER[1]

n 1539, a Polish doctor, economist, mathematician, and part-time astronomer named Nicolaus Copernicus published a paper he had been working on for some years. The paper explained a strange idea: The earth revolves around the sun.[2] His idea flew in the face of common sense. Everybody knew from direct evidence of their own eyes that the sun revolves around the earth. Each day, people from their motionless position on the firm solid earth saw the sun rise in the east, travel across the sky, and set in the west. It was obviously so. Copernicus was saying: "Forget common sense. Forget what your eyes are telling you. Forget what everyone believes. Common sense is wrong. The earth revolves around the sun."

When Copernicus first presented his idea, it seemed to be no more than a different mathematical model for astronomers and astrologers—a simpler way of calculating the paths of the planets. At the time, his sun-centric view of the world did not offer more accurate predictions of planetary positions than the earth-centric mental model of the world. But his theory appealed to astronomers and astrologers as a simpler and

49

clearer way of comprehending the heavens, as well as pointing to more fruitful hypotheses to explore.

Initially the powers-that-be were thrilled. When Pope Clement VII received an explanation of the theory in 1533 by his own secretary, Johann Widmannstetter, he was so pleased that he offered Copernicus a valuable gift.

What Pope Clement didn't grasp was that the Copernican theory wasn't just a mathematical model for calculating the movements of the planets. Embedded within it was a different worldview that implicitly undermined the plausibility of established religion in general, the Roman Catholic Church in particular, and the divine right of kings, on which most European governments based their claim to legitimacy.[3]

The publication of Copernicus's theory thus began not just a rethinking of astronomy (Figure 3-1), but an inexorable process of inquiry into the entire organization of society, including the rights and privileges of the individuals who happened to be in charge of the Roman Catholic Church and of the monarchies that asserted power by divine right.

It became possible to ask what social value those presiding over these institutions were adding. While some of them continued to be seen as genuinely wise and courageous leaders, others were discovered to be petty tyrants, plodding bureaucrats, or incompetent nincompoops. By stripping these people and organizations of their cosmic legitimacy, the Copernican theory enabled their true social worth to be examined and recalibrated. Royalty and churches continued to exist, but they occupied a steadily diminishing role in the structure of society.

Figure 3-1. The Copernican revolution in astronomy.

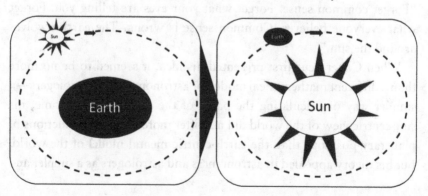

In due course, the powers-that-be grasped the gravity of the challenge. In March 1616, almost eighty years after the publication of Copernicus's thesis, the Roman Catholic Church issued a decree banning it until it could be corrected and forbidding the publication of any similar book. In 1633, the Italian astronomer Galileo Galilei was convicted of heresy for supporting the sun-centric view of the world and was placed under house arrest for the rest of his life.

But to no avail. The revolution that Copernicus had launched was under way. Resistance was futile, even though it continued for a long time. It wasn't until several centuries later—1822, to be precise—that the Church finally conceded defeat and lifted the prohibition on discussion of the revolution in astronomy.

"To describe the innovation initiated by Copernicus as the simple interchange of the position of the earth and sun," wrote Thomas Kuhn, "is to make a molehill out of a promontory in the development of human thought. If Copernicus' proposal had had no consequences outside astronomy, it would have been neither so long delayed nor so strenuously resisted."[4]

■

Under the Law of the Customer, the practice of management is currently undergoing an equivalent transformation in the way organizations understand, and interact with, the world. The earliest and simplest articulation of the law came from Peter Drucker in 1954: "There is only one valid definition of business purpose: to create a customer."[5]

Drucker wrote, "It is the customer who determines what a business is. For it is the customer, and he alone, who through being willing to pay for a good or for a service, converts economic resources into wealth, things into goods. What the business thinks it produces is not of first importance—especially not to the future of the business and to its success. What the customer thinks he is buying, what he considers 'value,' is decisive—it determines what a business is, what it produces and whether it will prosper."[6]

Drucker's proposition was a radical departure from the common sense of the day. Everyone knew that a business was in business to make money. It had been so since time immemorial. Any conversation with a businessman or economist confirmed that it was so. In fact, support for the view of money-making as the sole purpose of a business steadily gathered momentum in the latter part of the twentieth century, despite Drucker's insight. In the 1970s and 1980s, it even hardened into a formal economic doctrine—namely, that the purpose of a firm is to maximize shareholder value as reflected in the current stock price. In the 1990s, it was cemented in place by stock compensation for the C-suite.[7] As we will see in Chapter 8, despite the noxious short-termism generated by this doctrine, it became the dominant mantra of public companies, particularly in the United States, for the next several decades.

Meanwhile, changes in the marketplace since 1954 were steadily reinforcing the validity of Drucker's insight and the embryonic Law of the Customer. Deregulation, globalization, the emergence of knowledge work, and new technology all made inroads. Competition increased. The pace of change accelerated. Knowledge workers became central to generating the innovation needed to create and retain customers. To top it off, the Internet transformed everything.

An epochal change in the commercial center of gravity thus occurred: Power in the marketplace shifted from seller to buyer. Instead of the firm being the stable center of the commercial universe, now the customer was the center. For firms to be successful, customers had to be not only satisfied, they had to be *delighted*. New technological capabilities meant that it became possible to deliver instant, intimate, frictionless value at scale. Once that became *possible*, as shown by firms such as Apple, Amazon, and Google, it also became *necessary*. In effect, instant, intimate, frictionless value at scale became the new standard of corporate performance that customers began first to expect and then to insist on. This is a fundamental reversal in most people's understanding of how the world works. It means that firms have to be looking at, and interacting with, the world differently.

Instead of giant corporations standing as the solid center of the marketplace making money out of passive consumers who float by and are manipulated, the living customer with mercurial thoughts and feelings is now the center of the commercial universe. The customer is the sun, and organizations orbit around it. It's a Copernican revolution in management (Figure 3-2).

By and large, the twentieth-century corporation had proceeded on the basis that it made certain products and services that customers could be induced, through marketing and sales campaigns, to purchase.

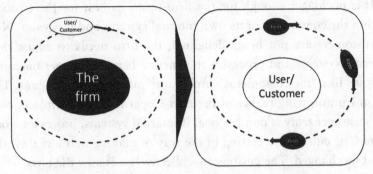

Figure 3-2. The Copernican revolution in management.

A car company made cars. A computer company made computers. A clothing firm made clothing. A newspaper firm produced a newspaper. If the firm performed well, the customers bought more. If it performed less well, the customers bought less. The purpose of the firm was to make money. That was the way the world was. The world had always been this way. The mindset was in effect: "We make it and you take it!" It was common sense. It was obvious.

A world in which the customer is the center of the commercial universe is a very different place. Customers now have options and reliable information as to what those options are. Customers can communicate with each other and share experiences. With social media, those experiences can reach vast numbers of other customers. Now the firm can't rely on telling customers to "take what we make."

With customers in charge of the marketplace, a firm-centric mindset has significant risks. Now the firm has to think about what the customers' problems are and try to figure out what might surprise and delight them by solving those problems. The firm needs to be looking at the world from the customer's perspective and understanding to what extent the outcome of interacting with the firm will improve the customer's life.

The shift implicit in the Law of the Customer, as Ranjay Gulati points out in *Reorganize for Resilience*, goes far beyond merely a fix to the existing management practices, such as upgrading the sales team or strengthening customer service.[8] It means reorienting everything in the organization to meet a different goal. It means mobilizing employees and partners to deliver more value to customers sooner, and aligning all communications, decision making, systems, structures, values, and culture with this goal.

It is no longer enough for the firm to do its best for the customer within the constraints of its own internal systems and processes. Now, if customers are not being delighted, the firm needs to adjust those internal systems and processes to generate better customer outcomes. Slogans like "the customer is number one" are no longer slogans. They are the unforgiving reality of the post-Copernican marketplace. Now, the customer truly *is* number one. If internal systems, processes, goals, values, or culture are getting in the way of making that a reality, they must be changed. The customer—collectively—is now the boss.

■

This is a massive challenge, with dramatic implications for how organizations are run and how society gets things done. We see certain characteristics in firms that have achieved post-Copernican performance:

1. There is a shared goal of delighting the customer.
2. Top management takes responsibility for ensuring enthusiasm for delighting the customer throughout the organization.
3. The firm aspires to be the best at what it does.
4. Everyone in the organization has a clear line to the customer.
5. The firm ensures it has accurate and thorough knowledge of the customer.
6. Staff members are empowered to make decisions.
7. The firm's structure changes with the marketplace.
8. Relationships are interactive, vertically, horizontally, internally, and externally.
9. Back-office functions are aligned to serve the customer.
10. Value for customers must be monetizable for the organization.

First, everyone in the organization is passionate about, and driven by, the goal of delighting customers. Everyone recognizes the performance requirement of instant, intimate, frictionless value at scale and understands how their work contributes to that performance. Thus, staff at firms like Riot Games, Spotify, and SRI International are thrilled to be creating and delivering cool products and services that delight customers. As Richard Sheridan at Menlo Innovations explains, employees experience joy in their work. Work becomes an energizing, ebullient, effervescent source of energy. Enthusiasm is restless, kinetic, and irrepressible. It is more than contentment: It leaps, bubbles, and over-flows. It is contagious. It moves ideas forward and catalyzes action.[9]

These workplaces have a very different feel from pre-Copernican workplaces, where only one in five workers is fully engaged in their work and almost as many are actively undermining the firm's goals.[10] In post-Copernican workplaces, people come to work with a spring in their step. Work becomes, as Noël Coward suggested, more fun than fun.[11]

The task of delighting customers is thus the job of everyone. It requires the efforts of everyone in the corporation—and beyond—to share insights and figure out ways to handle a challenge that is much more difficult than merely delivering a product or service. Everyone does their best to get inside the minds and lives of end-users and intuit what could improve their lives, sometimes even before the end-users themselves know it. Since those doing the work are now typically skilled professionals and have access to information through interactions with users, colleagues, and the Internet, they often know more about how to meet the challenge than managers. Yet managers also have a key role to play.

Second, instilling enthusiasm for delighting customers throughout the organization is a key responsibility of top management. The top sets the tone for the organization. The C-suite thinks, feels, lives, and breathes customer delight. It makes clear, both inside and outside the firm, that the firm is not just in business to make money for its shareholders and senior executives. The senior managers evince their commitment to generating customer delight on a daily basis. It is the intensity of the top management's passion for delighting customers that drives the firm's ambition and the vastness of its goals, along with the dedication and attention to the minute details necessary to deliver on those goals. It is the audacity of the top management's commitment to deliver customer delight, combined with a pragmatism to put it into action, that prompts awe, builds support, and drives action. It is in this way that the top management fulfills its responsibility to create both meaning *at work* and meaning *in the work* being done.[12]

The C-suite can't perform this function by reciting a clever speech crafted by the PR department. It must come from the heart. A "numbers guy" with no particular interest or background in the firm's products and services will have a hard time performing the top-management function in the post-Copernican organization. Faking passion won't work. The staff immediately sees through a false front or a PR façade.

The Law of the Customer is a single message, delivered consistently, every day, all the time. Every utterance and every decision reflects the aspiration for the firm to be the best in the world at what it does—delighting customers by delivering instant, intimate, frictionless value.

The communications and actions of the entire C-suite in a post-Copernican firm reflect this reality. The C-suite can't say one thing to staff, yet something different to customers, and then the opposite to Wall Street. The firm can't have the CEO inspiring the staff to delight the customer and then have the chief financial officer eliminating work that is key to generating customer delight, all the time reassuring Wall Street that the firm is totally focused on maximizing shareholder value as reflected in the current stock price.

Consistency is key. A single statement from the top management to the staff that is inconsistent with the goal of delighting customers can spark a major setback. If repeated, there is a risk that the culture will go into a tailspin and begin regressing back to bureaucratic management.

In his book *That Used to Be Us*, Tom Friedman referred to this situation as "Carlson's Law":

> *Innovation that happens from the top down tends to be orderly but dumb. Innovation that happens from the bottom up tends to be chaotic but smart.* This makes it all the more important for every worker to be a creative creator or creative server and for every boss to understand that the boss's job is . . . to find ways to inspire, enable, and unleash innovation from the bottom up, and then to edit, manage, and merge that innovation from the top down to produce goods, services, and concepts.[13]

The actions and communications of the top management are part of an organic process that is both bottom-up and top-down. A principal role of the top is to identify and support champions throughout the organization. If the drive to delight customers comes from the CEO alone, or from the bottom alone, the firm is lost. At the same time, it is the top management that is responsible for remedying any systemic shortfalls or setbacks in delivering customer delight.

Third, the firm aspires to be the best in the world at what it does. There is no second place. To have a sustainable future, the firm recognizes that it must be world-class. In a global economy driven by the

Internet, every business is potentially a global business, facing global competition that could come from anywhere in the world. It knows that customers expect the best, can find out what it is and who is delivering it, and thanks to the Web, can take steps to get it.

As a result, the firm does its utmost to recruit, nourish, and develop expertise that is the basis for being—and staying—the best. Outsourcing is both an opportunity and a trap. The firm may outsource aspects of its work where it's not the best. Yet, as we shall see in Chapter 10, outsourcing poses risks: Outsourcing may save costs in the short term, but the firm may lose expertise that limits its capacity to compete in future.

Fourth, everyone in the organization has a clear line of sight to the ultimate customer or end-user, which enables everyone to see how their work is adding value to customers—or not. If not, they can ask: Why are we doing this work at all? This line of sight is supported and clarified by performance measures, particularly measures used by the financial function, which enables day-to-day decision making to reflect the goal, at every level of the organization.

Fifth, the firm ensures that there is accurate knowledge of customers' context, goals, constraints, feelings, aspirations, and fears to be able to figure out what will delight them. As we saw at Menlo Innovations, anthropologists can be used to gather this understanding. Site visits and interactions with customers also add value. Recruiting customers as staff is a widespread phenomenon, as at Apple and Spotify. At Riot Games, being an enthusiastic gamer is almost a requirement of working there.

Yet firms also take care that staff members don't impose their preferences on the customers. Thus, although Spotify tends to recruit music aficionados, it also makes efforts to track and understand customers' preferences, which often differ markedly from the tastes of its own staff.

What's more, these firms keep learning *systematically*. Significantly, of the thirteen public companies that have outperformed the S&P 500 five years in a row, including Facebook, Amazon, Google, and Salesforce, most of them are algorithmically driven. Like Spotify (which is not yet public), they are not only gathering data from their users but also *systematically* learning from that data to update and enhance the user experience. They are using multiple sensors and people to gather information about the current outcome of customers and almost automatically adjusting and upgrading their products and services on the fly from the information they have gathered.[14]

Spotify's Discover Weekly is a good example. Depending on how you respond to the music playlist that Spotify sent you this week, next week's playlist will be adjusted accordingly. It's adjusted for you *individually,* as well as for each of the other more than 100 million Spotify users. Which is kind of amazing, and even a bit spooky.

The result is that brands are becoming less important than they used to be. (There's only one consumer company in the top thirteen public companies.) Users are becoming less interested in the brand—what the firm did for them yesterday—and steadily more interested in what the firm is offering today. Through the Internet, they have the means to find that out and make their decisions accordingly.[15]

Sixth, staff members are empowered to make decisions to enable a better outcome for the customer. Decisions are pushed to the lowest possible level in the organization—preferably to the small teams that make up the workforce. These teams are not waiting around for decisions or approvals from the top. They get on with whatever is necessary to delight customers. As we will see in Chapter 4, the role of a senior manager changes from being a fierce, decision-taking commander to something more like a curator or gardener.

Seventh, the firm's structure is a fluid and ever-changing network that reflects its customer focus in a rapidly changing marketplace. As the marketplace is constantly in flux, the organization itself is also constantly in flux. To ensure that outcomes for customers have primacy over internal unit preoccupations, the firm shies away from any static or permanent pyramidal structure. Any such structure will inevitably impose its own internal dynamic and preoccupations on the firm's products and services, ahead of needed customer outcomes.

Thus, the Law of the Customer requires that the firm's culture and internal systems, processes, and values themselves be continuously subordinated to, and driven by, delivering value to the customer: If there is a conflict, it is the customer's needs that need to be given priority. Because of the clear line of sight to the ultimate customer, everyone can understand how the work contributes value to the customer and can take steps to make any necessary adjustments. Instead of frequent and massive "reorganizations" that periodically disrupt the supposedly permanent structure of a bureaucratic pyramid, there is constant adjustment and flux at all levels.

Eighth, relationships are interactive, both vertically and horizontally, and both inside and outside the organization. Customers and partners become part of the organization as active participants in creating value. With the push for continuous innovation, there is a recognition that ideas can come from anywhere. In the ensuing interactions, the boundaries between the organization and its context tend to dissolve (as shown in Figure 1-2 in Chapter 1).

Ninth, back-office functions, such as finance, budgeting, accounting, auditing, procurement, and people management, are aligned to reflect the primacy of the customer. Unlike a bureaucracy, where back-office functions take on a life of their own, the Law of the Customer requires that all internal systems and processes themselves be subordinated to, and driven by, delivering value to the customer. If there is a conflict, it is the outcome for the customer that is given priority. As with operations, all these back-office functions have a clear line of sight to the ultimate customer and continually adapt their systems and processes to enhance their contribution to the outcome for the ultimate customer. If the systems and processes aren't so contributing, it becomes legitimate to ask: Why do these systems and processes exist at all?

Aligning the finance function with the Law of the Customer is a key priority. In public corporations, there is often a life-threatening struggle to liberate the firm from the pressures of rampaging short-termism from the stock market, as discussed in Chapters 8 to 10.

Recruitment of staff who share the goals, values, and attitudes of Agile management becomes a high priority to ensure that the customer-oriented culture is continuously strengthened, rather than undermined. Career development and compensation policies are adapted to reflect and support both the Law of the Small Team and the Law of the Customer. As an illustration of what the reconciliation of recruitment, training, and career development policies with Agile management at a large corporation looks like, the story of the Agile transformation at Cerner Corporation offers many insights (see Box 3-4).

Tenth, firms keep in mind that value for customers must in the end be monetizable for the organization. While committing to the primacy of the customer, the firm also recognizes that it must make money to survive. However, making money is the result, not the goal.

Happily, the Law of the Customer is well adapted to making money, with gains on both pricing and costs. On pricing, firms that delight their customers can have higher margins because the customers *must* have the products and services they love, and they are willing to queue up and pay extra for it.

Costs also tend to come down for several reasons. One is that firms stop doing things that customers don't care about and so save money. Instead of doing what's in some big plan, with items that the firm *thinks* customers want, the firm only does what customers *actually* want.

Another reason is that firms that delight their customers compete on time and get work done faster.[16] When work gets done faster in the effort to delight customers, costs tend to come down of their own accord. "Capitalizing on time [is] a critical source of competitive advantage," writes George Stalk. "Shortening the planning loop in the product development cycle, trimming process time in the factory, drastically reducing sales and distribution—managing time the way most companies manage costs, quality, or inventory. In fact, as a strategic weapon, time is the equivalent of money, productivity, quality, even innovation."[17]

Yet another reason costs tend to come down concerns savings in sales and marketing. The firm that has delighted customers also has the advantage of an unpaid marketing department. Its customers are thrilled to sing its praises to friends and colleagues, so the firm can sit back and watch. And the firm doesn't have to spend time and money dealing with disgruntled customers or counteracting negative social media blitzes. The firm does the right thing—delighting its customers—in the first place.

■

Taken as a whole, these elements add up to a revolution in management for most firms. As in astronomy, the Law of the Customer has thus begun not just technical adjustments to arcane management systems and processes, but also a wider inquiry into the way entire organizations are run, including the rights, privileges, and duties of the individuals who happen to oversee them.

As Thomas Kuhn might say, to describe the innovation initiated by the Law of the Customer as the simple interchange of the position of the customer and the shareholder would be to make a molehill out of a promontory in organizational thinking. As in the Copernican revolution in astronomy, if Agile management had no implications beyond the nitty-gritty of specific work processes, it would be neither so long delayed nor so strenuously resisted. Agile management begins with simple shifts in how work is done, but ends with questions such as: Which class of people will be running organizations? How much will they be compensated? How will the overall economy function and how will society as a whole get things done? In the end, it is about power.

As columnist Peggy Noonan wrote in the *Wall Street Journal*:

There is an arresting moment in Walter Isaacson's biography of Steve Jobs, in which Jobs speaks at length about his philosophy of business. He's at the end of his life and is summing things up. His mission, he says, was plain: to "build an enduring company where people were motivated to make great products." Then he turned to the rise and fall of various businesses. He has a theory about "why decline happens" at great companies: "The company does a great job, innovates and becomes a monopoly or close to it in some field, and then the quality of the product becomes less important. The company starts valuing the great salesman, because they're the ones who can move the needle on revenues." So salesmen are put in charge, and product engineers and designers feel demoted: Their efforts are no longer at the white-hot center of the company's daily life. They "turn off." IBM and Xerox, Jobs said, faltered in precisely this way. The salesmen who led the companies were smart and eloquent, but "they didn't know anything about the product." In the end this can doom a great company, because what consumers want is good products.[18]

And it's not just the salesmen. It's also the accountants and the money men who search the firm high and low to find ingenious ways to cut costs and extract value for shareholders.[19] These activities dispirit the creators, the product engineers, and the designers, and they crimp the firm's ability to add value to its customers. Yet because salespeople appear to be adding to the firm's short-term profitability and share price, as a class they are currently celebrated and rewarded, even as they are systematically undermining the firm's future.

Such firms are operating with pre-Copernican thinking and basically playing defense. The firm not only stops playing offense, it even forgets how to play offense, and the firm starts to die. If the firm is in a quasi-monopoly position, it may go on extracting value for extended periods of time. But basically, the firm is in a death spiral, as it dispirits those doing the work and frustrates its customers.

As managers playing solely defense find it steadily more difficult to make money, they tend to become more desperate and start doing ever more perilous things to extract value (as we will see in Chapters 8 to 11), like looting the firm's pension fund,[20] or cutting back on worker benefits,[21] or outsourcing production to a foreign country in ways that further destroy the firm's ability to innovate and compete.[22] When many firms are operating in this way, we have a whole economy operating in what economists call "secular economic stagnation." Inequality increases and the political and social fabric starts to unravel.[23]

■

Why do managers keep on a path that is out of sync with the VUCA marketplace and that is systematically killing their firms? For one thing, it's more difficult to create value than to extract value. For another, executives have found ways to reward themselves lavishly for extracting value. As Upton Sinclair noted long ago, "It is difficult to get a man to understand something, when his salary depends upon his not understanding it."[24]

It's also the case that genuine paradigm shifts, by their very nature, are always resisted. And let's be clear. We are talking about a paradigm shift in the strict sense as laid down by Thomas Kuhn: a different mental model of the world. The changes involved here go far beyond those business articles in which the term "paradigm shift" has been applied to some tweak to the existing managerial canon, such as a new negotiating tactic or a different HR practice. In that degraded sense, "paradigm shifts in management" have been claimed so frequently and inappropriately over the last half century in management writing that the very term is a standing management joke. Management writers have cried wolf so often that now, when a real wolf comes along, it is hard to recognize it as a wolf.

As it happens, the parallelism of paradigm shifts in science and management is striking, as we can see from Thomas Kuhn's pathbreaking book *The Structure of Scientific Revolutions* (which is summarized in Box 3-1).[25]

In the nineteenth century, management was in a *pre-paradigm phase, in which there was no consensus on any consistent theory of management.* Management thinking entered the *second phase* with the work of Frederick Taylor and his principles of "scientific management" in 1911, which began with the ominously prescient declaration: "In the past, Man has been first. In future, the system must be first."[26]

Despite many changes and evolutions, the system that Taylor initiated has several underlying assumptions that are still "obvious" to many managers and theorists today. They include running the organization through top-down command and control. The top knows best. Bureaucracy is the only way to run an efficient organization on a large scale. A corporation can manipulate customers to buy its products and services. The overriding objective of the firm is to make money for shareholders with ever-greater predictability and efficiency through economies of scale.

For the next hundred years or so, these assumptions, and the systems, values, and corporate cultures that are built on them, have remained the default mental model of management. As in science, most managers and theorists have spent their entire careers accepting the prevailing paradigm and proceeding in a puzzle-solving mode within these assumptions.

Yet over the last half century, many unresolved anomalies have emerged. There is the need to cope with a much faster pace of change, including unexpected developments in technology and the shift in power from seller to buyer. There is the need for continuous innovation, not just exploiting the existing business. There is the need to cope with the mercurial whims and wishes of customers who now have a decisive say in the future of the organization. There is the need to respond to the accelerating disruption of the existing business by upstarts that are moving faster and more responsively to meet the needs of customers, sometimes coming from other sectors. There is the need for more attention to motivation and the human dimension of work, particularly through teams and collaboration, in a marketplace where innovation is central. And so on.

Throughout the twentieth century, these "anomalies" were dealt with by grafting fixes onto the prevailing mental model of management without basically changing it. Firms tightened management control. They downsized. They reorganized. They delayered. They empowered their staff. They reengineered processes. They expanded sales and marketing campaigns. They acquired new companies. They shed businesses that weren't doing well. These fixes sometimes led to short-term gains, but they didn't solve the underlying problem.

In retrospect, the reasons are obvious. How could the firm be nimbler when decisions needed approvals up a steep chain of command? How could the firm pursue innovation when all its systems and processes were devoted to preserving the status quo and extracting the gains from efficiency? How could the firm encourage creativity and collaboration when the basic structure of work involved having bosses tell individuals what to do? How could trust and transparency be achieved when the vertical chains of command of a bureaucracy inherently fostered nontransparency? How could the firm innovate with a dispirited, even disruptive, workforce?

In practice, managers lurched backward and forward, one moment paying more attention to resolving the anomalies and then reverting back to the dominant mental model, particularly when it became apparent that attending to the anomalies endangered short-term returns to shareholders or put in question the managers' control.

■

Another aspect that delays paradigm shifts is when they come from the "wrong people."

Thus in 1714, the British government offered a prize for a method of determining longitude at sea, with an award of £20,000 (£3 million in today's terms). John Harrison, a Yorkshire carpenter, worked on the project for several decades and eventually, in 1761, came up with a design that proved accurate on a long voyage to Jamaica. The scientific establishment refused to believe that a Yorkshire carpenter could possibly have solved the problem that had stumped the best scientific minds. Some twelve years later in 1773, when Harrison was eighty years old, he received a monetary award in the amount of £8,750 from Parliament for his achievements, but he never received the actual prize. A Yorkshire carpenter was the wrong person to have solved the problem.

In 1865, an unknown professor named Gregor Mendel read a paper at two meetings of the Natural History Society of Brünn in Moravia, giving the results of studies in which he had cultivated and tested some 29,000 pea plants. His study presented a solution to a problem that had stumped the finest scientific minds. The paper was ignored by the international scientific community for the succeeding thirty-five years until it was eventually realized that Mendel had indeed come up with the solution. His work later became known as Mendel's Laws of Inheritance and he was hailed as the father of modern genetics. His work was ignored for decades by the scientific community because a researcher on peas in Moravia was the wrong person to have solved the problem.

In 1981, Barry Marshall, a pathologist in Perth, Australia, came up with an odd idea: Stomach ulcers are caused by the presence of spiral bacteria. His idea was at odds with the thinking of the international scientific community. It was ridiculed by the medical establishment. How could bacteria possibly live in the acidic environment of the stomach? In 1984, frustrated by the widespread disrespect for his ideas, Marshall drank a Petri dish of liquid containing the spiral bacteria, expecting to develop, perhaps years later, an ulcer. He was surprised when, only three days later, he developed ulcer symptoms. But it took two more decades before he was awarded the Nobel Prize for Medicine. An obscure pathologist from Perth was the wrong person to have solved the problem.

Over and over again, we see that when a bold new idea challenges an entire way of thinking of the establishment, if the idea comes from an unexpected source, it can languish in obscurity for decades, even though the solution to a problem that needs to be solved is staring the experts in the face.

Something similar is occurring now in management. Over the last decade, the Law of the Small Team and the Law of the Customer have been field-tested in thousands of organizations around the world. Unfortunately, these management discoveries weren't made by "the right people"—academics in suits and ties at business schools or high-paid consultants. The discoveries were made by the people you might think are the least likely people to have solved a management problem: geeks. Software developers were known to be antipathetic to both managers and management. Often badly dressed and disrespectful, they were often seen as the most problematic of a big organization's employees. It was preposterous to think that innovation in management could come from such people.

Yet in retrospect, it's not hard to see why software developers would be the ones to solve the problem. In the 1990s, huge sums of money were being lost because the work of software development was always late, over budget, and plagued by quality problems. Clients were upset, and firms lost money. Developers were seen as culprits and were punished. They worked harder and harder. They labored evenings and weekends. It made no difference. The software was still late, over budget, and full of bugs. They were fired, but their replacements did no better. The standard prescriptions of management didn't work with software development. It was frustrating for managers to find that the more they tried to control things, the less progress they made. Complexity responded to competence, not authority. Something different *had* to be found.

■

Making the shift to the new paradigm isn't easy. It's one thing to accept intellectually that adding value to customers is the purpose of an organization. It's another to take the goals, systems, processes, structures, values, habits, attitudes, and culture of an entire bureaucracy and turn that intellectual principle into a reality.

Given the deep change involved, moving from the pre-Copernican to the post-Copernican world of management doesn't happen overnight. When there are only a few teams, the organization might learn to operate routinely in this way within a year. When there are many teams, managing dependencies between the teams becomes more of a challenge and it may take several years. With a large organization, it will take many years for these principles to become second nature to the organization. There are many new behaviors, attitudes, and assumptions to learn and internalize so that they become "the way we do things around here." There are also many processes in people management, budgeting, accounting, and auditing that will also need to be aligned with the new way of working.

The full gains from post-Copernican management come when the whole organization is operating under the Law of the Small Team and the Law of the Customer. How is this possible? It is to this set of issues that we turn in the next chapter with the Law of the Network.

BOX 3-1

PARADIGM SHIFTS IN SCIENCE

The parallelism of paradigm shifts in science and management is exact. Let's refresh our memories of Thomas Kuhn's *The Structure of Scientific Revolutions* and remind ourselves: What is a paradigm shift in science?

In 1962, Kuhn's book challenged the prevailing view of progress in science. Until then, progress in science was seen primarily as the steady accretion of new facts and relationships, one on top of the other. Kuhn accepted that science experiences long periods of conceptual continuity and accretion. This is what he calls "normal science." However, it is punctuated by periods of "revolutionary science," with abrupt discontinuities as the mental model changes in fundamental ways. There are three distinct phases.

The first phase is the *pre-paradigm phase*, in which there is no consensus on any particular theory. There may be several incompatible or incomplete theories.

The second phase occurs when a number of puzzles can be solved within a single mental framework or paradigm. This then becomes the *dominant framework*. Scientists gravitate to the new framework and normal science begins. Thereafter, scientists try to solve puzzles within the assumptions of the dominant paradigm. Most scientists spend their entire careers accepting the prevailing framework and proceeding in a puzzle-solving mode within its basic assumptions. It becomes unthinkable to view the world in any other way.

As time goes on, however, anomalies appear that cannot easily be resolved within the dominant framework of assumptions. Yet paradoxically, puzzle-solving within the framework continues with even greater tenacity, because prior successes lead people to believe that resolving the anomalies within the existing framework *must* be possible, even though solutions are hard to find. Anomalies thus accumulate as the dominant paradigm is stretched and bent and adjusted in an increasingly desperate effort to accommodate them, yet with only partial success.

The third phase occurs when some scientists finally accept that the dominant framework can't resolve the anomalies. These scientists come to see that the basic framework can't be repaired. It must be replaced. They start exploring alternatives to long-held, seemingly obvious, self-evident assumptions.

And thus, *revolutionary science* begins. These scientists develop a *new conceptual framework* that they believe presents a better way of reconciling the known facts with the anomalies. At first, the new framework itself has various gaps and contradictions, because it is incomplete and needs further thought.

When first presented, the new framework is greeted with hostility and derision from the establishment—people whose careers and lives have been based on the nurturing of the existing paradigm. The inevitable gaps and flaws in the new framework are cited as reasons for not adopting it. Consequently, paradigm shifts do not occur quickly or easily.

There follows a period in which the adherents of different frameworks pursue their different theories. The powers-that-be attack the revolutionary framework for being theoretically unsound, incomplete, and irresponsible, while the revolutionaries point to the growing list of unresolved anomalies of the still-dominant paradigm.

In time, the gaps in the new framework are filled and the new framework is integrated and completed. Other scientists then converge on it as a more productive way of understanding how the world works. Once most scientists agree that the new framework should replace the old one, a paradigm shift has occurred, even though many individuals may remain intransigent and continue to think and teach in the old way.

As scientists look at the world in the new way, the direction of future scientific research in the field shifts. New questions are asked of old data. In due course, textbooks are rewritten and university courses are revised.

The period of conflict between frameworks may last for decades, particularly if the new paradigm comes not from some of the existing leaders of the current scientific establishment, but rather from some unexpected and peripheral source.

BOX 3-2

ULTIMATE CUSTOMERS, INTERNAL CUSTOMERS, AND END-USERS

In this chapter, I use the word "customer" and "user" as synonyms. That reflects the simplest case in which the ultimate customer is also the end-user, such as when someone buys an iPhone. (See Figure 3-3.)

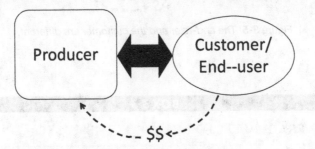

Figure 3-3. The end-user is the customer.

In some situations, the producer may be providing products or services to an internal customer who in due course becomes a producer for the ultimate customer. (See Figure 3-4.) In bureaucracies, the needs of the internal customer are often disconnected from the needs of the ultimate customer, resulting in waste. Under the Law of the Customer, the original producers not only meet the needs of the internal customers, they have a clear line of sight as to what value is being provided for the ultimate customer.

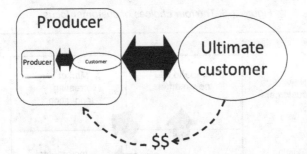

Figure 3-4. Internal and ultimate customers.

There are also situations where the end-user and the customer are different. Those using the Google Search capability—the end-users—get it for free. The paying customers are the advertisers: Google must satisfy both customers and end-users (see Figure 3-5).

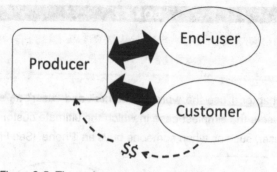

Figure 3-5. The end-user and the customer are different.

BOX 3-3

PRACTICES OF THE LAW OF THE CUSTOMER

For firms that have truly made the shift to the customer-driven mindset, here are some of the practices that tend to emerge.

1. **Target.** Identify your core market of primary customers. Delighting this group is important so that you have a resilient customer base. Trying to satisfy everyone at the outset practically guarantees average products and services that will not delight anyone. Careful choices need to be made in terms of where to put one's efforts.

Figure 3-6. The four choices of customer focus.

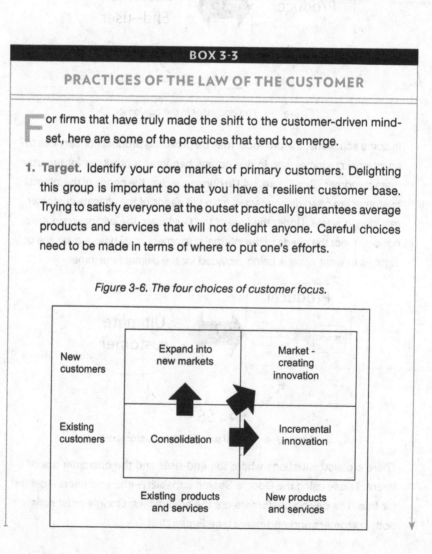

The bottom-left quadrant in Figure 3-6—doing what the firm has always done—was once the safest place. Now by itself, it has become the most dangerous—preserving the status quo at the very time that your customers are likely to be thinking about new options being offered by competitors.

The bottom-right quadrant—incremental innovation by cost reduction or quality improvements—is necessary. It's the price of survival. But it may not be enough by itself to generate major new revenue gains.

Huge new profits will generally have to come by moving into the top-left quadrant (new markets for existing products and services) or the top-right quadrant (generating new products and services that create new markets). We'll come back to the top-right quadrant in Chapters 6 and 7 as it requires a whole new level of agility: going beyond operational Agility to Strategic Agility.

2. **Constantly experiment.** "In this environment," says management consultant Paul Nunes, "you are better off to experiment constantly and in the market, learning and sensing, and creating a wealth of options, rather than trying to determine 'best' a priori, at least so long as you can afford to do that. In industries like software, where the cost of experimentation and revision is low, that is truer than some others. Sites like Kickstarter have demonstrated how you can experiment with offers of all sorts of products that you haven't even created. That's why we say market experimentation is necessary. That doesn't mean, however, you don't have to have a strategy, or can't win with a single shot. By learning from the experiments of others, and timing cost trends perfectly, you can create a Big Bang Disruption like Kindle. So, there are different ways of discovering and creating the learnings of experimentation, and of creating social consensus."[1]

3. **Partner with startups.** Don't try to do everything yourself. There may be some parts of the firm's DNA that get in the way of what the firm needs to do. Take, for instance, videos on *The New Yorker* website. The writing in that magazine is superb—arguably the very best in the world today. But their videos? Not just bad: deplorable. It seems that *The New Yorker* as an organization has a blind spot for video. They should accept that reality and partner with a startup that has video in its DNA.

But don't just bolt the startup onto the bureaucracy. Embrace the startup way of operating for the whole firm.

4. **Increase product malleability.** To the extent possible, shift from hardware to software so that the product can be easily upgraded and customized. But also recognize that "going digital" is not by itself the answer. Merely turning a physical product into a digital product will achieve nothing unless it is accompanied by a shift in mindset to see things from the customers' perspective and generate better outcomes.

5. **Focus.** Aim for the simplest possible thing that will delight buyers. Don't load products down with features that most people won't use and that make the product hard to operate. For instance, my DVD controller made by Sony has fifty-four buttons, most of which no one in our house knows how to use: It delights no one. By contrast, my iPod made by Apple has just four buttons and delights everyone.

6. **Innovate in short stages.** Launch the product or service with the key features that primary clients want, and then add selectively through upgrades. Apple's iPhone initially lacked many features of existing smartphones, but it delighted its core group of customers—young users who wanted a cool mobile phone; the other features were added later.

7. **Evaluate.** Don't just add features. Following every customer suggestion can lead to a client-driven death spiral. As more and more customer requests are met and features are added, the product can become unlovable or even unusable. Make sure that each upgrade really does delight.

8. **Be willing to disappoint.** Know *who* you want to delight. Southwest Airlines does not offer hot meals. Amazon does not delight book authors, publishers, or brick-and-mortar retail product sellers.

9. **Deliver value faster.** It must be as close to instant as possible. For customers, time is often the most valuable resource. In a bureaucracy, time is constantly wasted, as work sits in queues, awaiting management approvals. Use value stream mapping to eliminate such delays.

10. **Customize.** Today's performance criteria include personalized products and services. Thus Harley-Davidson isn't merely building reliable motorcycles. It aims to fulfill the dreams of its customers through the motorcycle experience. If that means going beyond the signature full-throated roar of their Harley and enabling Harley owners to embellish their vehicles with grassroots folk art, the company will help them do it.

NOTE

1. P. Nunes, quoted in S. Denning, "The Business Disease Without a Cure: Big-Bang Disruption," *Forbes.com*, February 22, 2014, https://www.forbes.com/sites/stevedenning/2014/02/22/ the-business-disease-without-a-cure-big-bang-disruption/.

BOX 3-4

ALIGNING PEOPLE MANAGEMENT WITH AGILE MANAGEMENT AT CERNER

A firm that has made considerable progress in creating a network culture is Cerner Corporation, the medical software firm, headquartered in Kansas City, Missouri, with around 25,000 employees and growing rapidly. Its Agile transformation began in 2008 when a massive effort was made to imbue the existing staff with the Law of the Small Team and the Law of the Customer.

With rapid growth, Cerner has a constant influx of new staff who do not necessarily arrive at Cerner with an Agile mindset. How does Cerner go about sustaining the elements of an Agile culture in a situation of constant flux? How does that work in the highly regulated health sector, where the risk of failure is a matter of life and death? How does Cerner inculcate and sustain a spirit of autonomy and innovation with reliability?

After Cerner had launched its Agile transformation, the firm found it wasn't easy to recruit people who could operate reliably in an Agile fashion, producing quality code, troubleshooting problems quickly and accurately, using automated testing, and seeing things from the end-user's perspective. It shouldn't have been surprising that new recruits weren't skilled in those areas because the colleges they recruited from weren't teaching those things.

Enter Michelle Brush, an executive within the Population Health business at Cerner. Her primary duties include developing solutions that identify those who are at risk for high-cost, low-quality-of-life blood conditions like diabetes or heart failure. Identification then leads to recommending interventions to be taken before patients get to the point that they need something serious like a foot amputation for diabetics. The goal is to catch problems through analytics and intelligence before it is too late. This results in improved health of the individual and significant reduced cost to the health care system.

Brush has an additional passion: improving the Agile culture at Cerner and nurturing it from the day an engineer first learns of Cerner. Under this umbrella, she has been responsible for the redesign of the engineering training and onboarding programs at Cerner.

The Problems with Training at Cerner

Prior to adopting an Agile development model, Cerner's engineering training focused on Cerner-specific technology and how Cerner interprets the health care domain. It was a four-week and a one-size-fits-all program. There were some assessments, including written tests followed by feedback. Yet Brush discovered that Cerner's managers were not seeing the desired behaviors after participants left the training program. So she set out to redesign the program with an Agile mindset and made it her personal challenge to solve.

These days, Brush talks as an executive. But she grew up through the ranks. Most of the associates who are involved in culture at the executive level now have that same story. They got involved at the very bottom, as individual contributors, and started proposing ideas. They got support from the leadership to drive those ideas and then they were rewarded and recognized for the changes they had made in Cerner's work environment. Now they have become executive champions. Her small team is always seeking out the next layer of associates who can come up and drive the culture forward.

In rethinking the concept of training, Brush found that Cerner was hiring associates who didn't really know how to do their work in an Agile fashion. That may sound surprising since there was already an elaborate interview process in place. And yet there were new associates who hardly knew how to start. This represented a small but significant percentage of new hires.

It was very expensive to give those associates four weeks of training, put them on a team, have the team work with them for six months, and then have the team come back and say, "We're sorry but it's not working. We can't coach them through this. They're not going to be a fit with Cerner."

Realizing that, Brush started to rethink the whole concept of training. She met with some sixty managers across Cerner, making sure to get the full diversity of opinions, experience, and roles. She asked what their principal challenges were working with candidates coming out of college. She built a list of all their concerns and the common patterns.

She found that the training program was built around teaching the Cerner way of doing things, the Cerner domain, and health care problems. However, the managers didn't care about that in respect of these recruits. What the managers wanted was something more practical. They wanted staff who knew how to troubleshoot, how to do automated testing, and how to talk to solution designers and ask the right questions to clarify user requirements. They wanted employees who were, in effect, living the Law of the Customer and the Law of the Small Team. There were too many new recruits who either thought you wrote code and just threw it over the wall or, at the other extreme, thought you had to fully document everything even before you started the project.

The Dev Academy

Brush realized that what was missing was behaviors. Cerner had to rethink what was meant by learning. The training program had been based on transferring knowledge. Instead, Cerner needed to focus on encouraging the right behaviors. Knowledge was not enough. Knowledge was just a means to getting the wanted behaviors.

Brush developed an inventory of the core behaviors that candidates weren't exhibiting and that were needed. She went through the data and threw out almost all the old training program. She took an Agile approach to developing the new program. They would roll out one new thing a week and get feedback. They would either keep it, tweak it, or throw it away. They did that over the course of six months until they finally had the kind of training program they wanted, which became known as the Dev Academy.

It begins with what is called "Dev Essentials": a fixed two-week program that introduces the wanted behaviors. This training includes DevOps, unit testing, Agile development, and some basic and general technical skills that everyone needs to have. Thus, everyone should know how to interact with the data store and what web development should look like.

Then the trainees receive a hands-on assessment. They are told which areas they need to work on that are essential to being an effective engineer at Cerner. It is very much like a classroom or college environment. They are teaching associates things that they might not have covered in college.

The Dev Center

After associates have completed the Dev Essentials, they are brought into the Dev Center and asked to pick a project. They receive a list of some forty real projects that are available at the time. They rank their preferences. There is a matching algorithm that helps put them into tiny teams of two or three.

The projects are real projects that Cerner needs to do anyway, although they are low-priority, low-risk projects. Cerner doesn't want the trainees touching clinical software. Clinical software is high risk: Cerner doesn't put that in production with brand-new associates. Instead, they are given projects like working on tools to facilitate troubleshooting or testing. All of the projects have business value, but they are very low risk.

Mentors help with the design decisions and with code. They give feedback and talk to the trainees about communication skills and what is expected from them in terms of quality. They spend about four hours a week being a mentor.

To supplement that mentorship, there are a handful of full-time staff at the Dev Center. These individuals are typically up-and-coming leaders in the company who help run the Dev Center for an eighteen- to twenty-four-month rotation. The result is that they come out of the Dev Center with a leadership position and essentially get to pick where they want to go.

In the Dev Center, the trainees pick their project, work on it, and get feedback. The mentors are required to say every week how the trainees are doing. The mentors fill out a very lightweight survey in which they are asked to look at fixed key performance areas. They are given guidelines on how to score in each area.

"We don't give [trainees] grades like A, B, or C," says Brush, "since people just out of school tend to freak out about grades. If a trainee receives a C, they are devastated. But if they receive a minus, they feel, 'Oh, I just need to improve a bit.' We need to get their thinking about grades out of their system."

Trainees get this feedback from the mentors weekly. They are told, "These are the areas for improvement." It might be more attention to quality or to unit-test coverage. The mentors monitor the trainees' performance weekly and hope to see the minuses turn into pluses or plus-pluses. It is entirely performance-based. When all of their ratings are pluses, they are assigned to a team—not before, and not later.

"This is different from the old program, which was very much fixed," says Brush. "It didn't matter how good or bad you were; you stayed exactly four weeks before you went to a team. Now a trainee can get out in two weeks or can be there for twelve weeks. It depends on what they need. It is run as a safe place. Associates aren't punished for taking twelve weeks. If they meet competency, they meet competency. They are going to do well at Cerner."

If the staff see signs that are concerning—for instance, that the trainees are lacking aptitude, or if they have been working for twelve weeks on one thing and they just can't get it, or where there is an inability to take feedback—then something more is needed. These associates are let go and they never impact a working team. "These days, not many associates are let go," says Brush, "which speaks to improvements that we have also made in the interview process."

The Dev Academy program is heavily documented, heavily audited, and heavily monitored. Everything is transparent and visible to everyone. Everyone can see who is in the Dev Center, what they are working on, and who their mentors are, although the performance reviews themselves remain private.

Improvements to Recruitment

When the program was reviewed after the first year, there was an issue whether to adjust the training program or make adjustments to hiring. Although there were some adjustments made to the training, they ended up making more adjustments to recruitment.

"That's because when we looked at the data," says Brush, "we saw a clear correlation between how Cerner hired and who the interviewer was, and how associates performed in the program. Based on the data, we launched an initiative to evaluate the hiring practices against research findings in psychology, which suggested once again a sharper focus on behaviors. It turned out that the interview process at the time was focused mainly on knowledge."

What candidates had done was seen to be a better predictor of what they would do in future than what they say they will do. So, Cerner decided to focus on behavior and refocused the questions on things like: "Tell me about projects you have worked on. Tell me about the most difficult bug you ever had to troubleshoot. How did you track it down? What was the problem? How did you fix it?"

The intent was to make it as much as possible an interesting conversation between the interviewer and the interviewee, rather than a quiz. It was like: "Let's talk about this! Let's go to the whiteboard together and discuss! Let's draw this out."

The candidate would be given situational questions to complement the behavioral questions. They would be given problems that were like real Cerner problems. For example: "Hey, we've got people coming into a hospital and we need to know who they are and where they live and what gender they are. Can you write me a data structure to gather that information?" They would critique the candidate in the process and make suggestions, like: "Why is that a string and not an object?" They are poking at the candidate's work as they are doing it. The purpose is to see how they respond to feedback, more than whether they have the right answer to the question.

As the interview goes on, the interviewers make the problem increasingly challenging until they get to the point where candidates are no longer able to respond to the problem. The interviewers don't hold the interviewees accountable for the point where they end. They are more interested in seeing how they respond to questions and whether they are able to incorporate new information as they talk through a problem. Are they able to rework their previous design to account for the new requirement that has just been introduced? The interview is meant to feel like two people sitting in front of a whiteboard, talking through ideas in a process of mutual discovery.

Cerner now has a standard interview packet with a pool of questions that the interviewer can pick. They don't have to ask all the questions. They ask them until they feel they have figured out the behaviors and the culture that they want. But they can't go outside the packet. Cerner insists on that because of evidence that consistent rubrics beat personal opinions every time. That's because people have unconscious biases. When interviewers say people "are not a good fit," what they generally mean is, "They are not right for me." To deal with this, Cerner insists on consistent rubrics and consistent approaches to interviewing.

For each question, the behaviors that the interviewer has to gauge are indicated. "What's their attention to detail? How do they handle and respond to feedback? How do they approach design?" What a good candidate, or an intermediate candidate, should have demonstrated by the end of the discussion of the question is also indicated. For each question, the interviewer circles the performance rating and this becomes the basis for discussion in the debriefing session.

Cerner is interested in engineering mobility. It is more important to know whether candidates will be able to learn software languages like Java or Ruby rather than whether they already know Java or Ruby. Technology changes so quickly, Cerner is constantly adapting. Cerner cares less about knowledge and more about the ability to incorporate new ideas, new thoughts, and new technology.

Cerner uses a pool of interviewers. They are not allowed to interview until they have gone through an interviewer training process. The recruiting organization regularly evaluates the quality of its interviewers, including how many interviewees accept offers and how their candidates work out over time. Cerner is frequently removing interviewers from the pool, based on the results and the feedback received. (For a time, Cerner tried getting away from the pool concept, but ended up going back to it. In the end, it is a matter of ensuring that there is an adequate inflow of good candidates.)

The focus is initially on "culture fit." The question of "team fit" comes in when the manager meets the applicant at the end of the process. A manager might say, "No one here fits my need." That means that the manager is opting to wait two weeks and get someone from the next batch. So, the culture fit is the basis of the decision to hire candidates into the pool. "Team fit" doesn't figure in the hiring decision and is less of a concern because Cerner doesn't emphasize keeping teams together for long periods anyway. It's more important that candidates can work on multiple teams.

Open Source

One thing that came out of the culture work at Cerner was a decision to enhance understanding of Cerner's place in the open-source community. While Cerner already used open-source methods, a review showed that Cerner was consuming more than contributing, let alone creating. Cerner recognized that as a creator of open-source code it needed to be creating code repositories all the time.

Cerner decided to take steps to get associates to understand the importance of open source. An outside speaker was brought in to help associates understand that contributing to open source is a competitive advantage for Cerner. By putting code out there for other firms to use, Cerner gets to define where the industry should go for that specific technology.

For instance, Cerner is now working on how to do streaming in big data sets. It's something that the industry hasn't yet solved. So the conversations have been about how to build something for Cerner that could solve the problem. Cerner wants to make it open source because it doesn't want to build something proprietary and then two years from now discover that the industry has come up with a different solution in open source and be forced to shift all its work to a solution that has become the industry standard. If Cerner is going to build something, it wants its product to become the industry standard. Or if there is something already out there in open source, Cerner wants to consume that and contribute to it. Cerner needs to throw its weight behind the solution that it believes is going to be the winning approach across the open-source industry.

Now Cerner is pushing associates to contribute to open source as much as they can, either through bug fixes or by drawing on open-source code in areas that are not Cerner's core competence. Cerner has also become an Apache Software Foundation sponsor and an organizational member of the public GitHub. When associates do contribute to open source, they are celebrated in town hall meetings or are otherwise acknowledged.

4.

THE LAW OF THE NETWORK

How does one build a team with seven thousand swim buddies?

—GENERAL STANLEY MCCHRYSTAL

I t wasn't a fair fight. The U.S. Army's Task Force in Iraq in late 2003 was arguably the world's most sophisticated fighting machine with abundant resources and advanced technology. It was facing a poorly resourced, ill-educated bunch of extremists. So why was the U.S. Army losing?

The Task Force was led by one of the army's star commanders, General Stanley McChrystal, who took charge in October 2003. He had at his disposal unmatched firepower, armored vehicles, cutting-edge surveillance, and new technologies such as precision weapons, GPS, and night vision. Descending from blacked-out helicopters that could locate a specific rooftop in a sea of buildings with pinpoint accuracy, operators communicated via headsets with pilots controlling drones that provided constant video surveillance.

The Task Force had achieved "the holy grail of military operations: near-perfect 'situational awareness,'" writes McChrystal in his book *Team of Teams*. "This was the first war in which we could see all of our operations unfolding in real time. Video feeds from unmanned aerial vehicles (UAVs or drones) gave us live footage of missions, while micro-phones carried by our operators provided audio. We enjoyed access to data on population, economic activity, oil exports, generation of electricity, and attitudes (through polling); we were connected to our partner organizations in real time."[1]

Before coming to Iraq, the Task Force's teams had several decades of successfully executed, precise, surgical operations. They had received the most rigorous training in the history of special operations. For sev-eral decades, the U.S. Army had been preparing for exactly the kind of challenge they were now facing in Iraq: asymmetric warfare. By any standard, these were high-performing teams, strong on trust and bonded together by common purpose.

On paper, the rabble of frustrated Sunnis led by Jordanian extremist Abu Musab al Zarqawi had no chance. They were ill-educated, ill-trained, underresourced, poorly equipped, and making do with primi-tive homemade weapons assembled in safe-house basements from pro-pane tanks and expired Soviet mortars. They had no coherent plan. They were making it up on the fly. They depended on face-to-face meetings and hand-carried letters. They were dogmatic and extreme in their conduct and views. They weren't military geniuses or tactical masterminds.

It wasn't just that the Task Force should have been winning this engagement. By every traditional military calculus, it should have been crushing its enemy. But it wasn't. The Task Force was losing badly and consistently and the situation was deteriorating.

At first, McChrystal couldn't understand how a poorly organized bunch of misfits could be defeating his crack force. It took him time to figure out the problem with his awesome military machine: The Task Force was a machine, while the enemy was operating as a flexible network. And in a turbulent environment with instant and pervasive communications, a machine—even a big, sophisticated, well-funded machine—is no match for a network.

The Task Force had ample access to big data, but that wasn't much help in terms of prediction. The Task Force was just too slow. McChrystal found himself being asked to make decisions and give approvals on matters on which the teams themselves were better placed to make the call. The decision-making apparatus was simply slowing down the ability of the Task Force to move as expeditiously as it needed to. He asked himself: Why am I being asked to make these decisions? What am I contributing?

"In the time it took to move a plan from creation to approval," writes McChrystal, "the battlefield for which the plan had been devised had changed. By the time it could be implemented, the plan—however ingenious in its initial design—was often irrelevant. We could not predict where the enemy would strike, and we could not respond fast enough when they did."[2]

McChrystal could also see that the problem wasn't collaboration *within* the teams themselves, but rather collaboration *between* the teams: "The bonds within squads are fundamentally different from those between squads or other units. In the words of one of our SEALs, 'The squad is the point at which everyone else sucks. That other squadron sucks, the other SEAL teams suck, and our Army counterparts definitely suck.' Of course, every other squad thought the same thing."[3]

The teams "had very provincial definitions of purpose: completing a mission or finishing intel analysis, rather than defeating [the enemy]. To each unit, the piece of the war that really mattered was the piece inside their box on the org chart; they were fighting their own fights in their own silos. The specialization that allowed for breathtaking efficiency became a liability in the face of the unpredictability of the real world."[4]

McChrystal could see that his superb teams were embedded in an authority-based bureaucracy in which communications and decisions flowed slowly and vertically. "Stratification and silos were hardwired throughout the Task Force," he writes. "Although all our units resided on the same compound, most lived with their 'kind,' some used different gyms, units controlled access to their planning areas, and each tribe had its own brand of standoffish superiority complex. Resources were shared reluctantly. Our forces lived a proximate but largely parallel existence."[5]

Communications, writes McChrystal, were "even worse between the Task Force and our partner organizations: the CIA, FBI, NSA, and conventional military units with whom we had to coordinate operations. Initially, representatives from these organizations lived in separate trailers, with limited access to our compound. Built in the name of security, these physical walls prevented routine interaction and produced misinformation and mistrust. The NSA [National Security Agency], for instance, initially refused to provide us with raw signal intercepts, insisting that they had to process their intelligence and send us summaries, often a process of several days. They weren't being intentionally difficult; their internal doctrine held that only they could effectively interpret their collections. Passing out raw data invited misinterpretation with potentially disastrous consequences."[6]

McChrystal knew he had to do something different. But what? He knew that "small teams are effective in large part because they are small—people know each other intimately and have clocked hundreds of hours with each other. In large organizations, most people will inevitably be strangers to one another. In fact," he writes, "the very traits that make teams great can often work to prevent their coherence into a broader whole. How does one build a team with seven thousand swim buddies?"[7]

The Task Force was too big to turn into one big team, yet it could not leave each team to its own devices. It needed rapid coordination of both information flow and collaborative action across the entire enterprise. The challenge was to achieve trust and purpose in a large group of teams and agencies without creating chaos.[8]

McChrystal knew he had to eliminate "the deeply rooted system of secrecy, clearances, and interforce rivalries." He had to reverse the principle of limiting information on a "need to know" basis to one where "everyone knows everything" so that "every man and woman in our command understood his or her role within the complex system that represented all of our undertakings. Everyone needed to be intimately familiar with every branch of the organization, and personally invested in the outcome."[9]

He could see that the ability to adapt to complexity and continuous unpredictable change was more important than carefully prepared plans. Rapid horizontal communications were more important than vertical consultations and approvals. Individual team excellence was not enough: Collaboration among teams was vital to achieve overall group performance. Teams had to be able to make decisions as needed, without seeking approvals higher up the command. Big data would never offer respite from the unrelenting need for continuous adaptability.

■

McChrystal's approach was a very simple idea: Create a team of teams. This meant turning the Task Force from a bureaucracy into a network (see Figure 4-1). A network is a group or system of interconnected people or things. An organizational network is a set of teams that interact with and collaborate with other teams with the same connectivity, interaction, and passion as they do within their own small team. An organizational network is founded on the Law of the Small Team. But it requires more. Each team needs to look beyond its own goals and concerns and see its work as part of the larger mission of the collectivity.

In effect, McChrystal had to transform the preexisting hierarchy of authority into a hierarchy of competence. Decisions had to be based on who was best placed to make the decision, not their position in the formal hierarchy. McChrystal had to find ways not only "to reverse the information flow—to ensure that when the bottom spoke the top listened, but also "breach the vertical walls separating the divisions of our enterprise. Interdependence meant that silos were no longer an accurate reflection of the environment. Events happening all over were now relevant to everyone."[10]

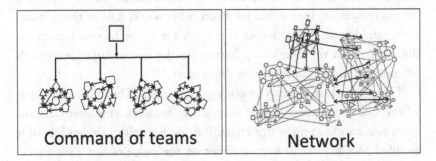

Figure 4-1. The difference between command of teams and a network.

Source: S. McChrystal, T. Collins, D. Silverman, and C. Fussell, *Team of Teams: New Rules of Engagement for a Complex World* (New York: Penguin Publishing Group, 2015), p. 24.

Embracing the seeming chaos of a network was a difficult personal decision for McChrystal to take. For centuries, military organizations had been built on this vertical and horizontal stratification. Although progress had been made in empowering individual teams to take initiative on the battlefield, the overall top-down command structure was still in place in the basic military structure.[11]

How did he transform an authority-based pyramid to a competence-based network?

First, it meant bringing all the key actors together in a *common physical space* to enable horizontal information flows, promote deeper understanding, and encourage initiative and decision making. "How people behave is often a by-product of how we set up physical space. At Balad [the makeshift command center], we needed a space that facilitated not the orderly, machinelike flow of paperwork, but the erratic, networked flow of ideas—an architecture designed not for separation, but for the merging of worlds. We weren't the only ones to be trying this—there was a growing movement in the private sector to organize offices for better cooperation, too," McChrystal writes.[12]

As he further describes this open space, "Anyone in the room—regardless of their position in the org charts, silos, and tiers—could glance up at the screens and know instantly about major factors affecting our mission at that moment. Personnel were placed strategically throughout the space, depending on their function—those with access to real-time information critical to ongoing operations were closer to the center of the room, those with a longer-term focus were on the fringes, so they could focus on other work. Any of them, however, could walk freely across the room for quick face-to-face coordination. With the touch of a button on the microphone, everyone's attention could be captured simultaneously."[13]

This wasn't just about symbolic egalitarianism. The cultivated chaos of the open space encouraged interaction between employees distant from one another on the org chart. Putting himself in the middle of it enabled McChrystal to keep a finger on the pulse of the Task Force. The open space enabled serendipitous encounters, with the recognition that no one could know in advance what connections and conversations would prove valuable.

Second, McChrystal himself embodied open communications in a *daily briefing session* that lasted an hour or two and in which everyone could understand how the overall operation was fitting together. It was an evolution of the daily standup (described in Figure 2-1, Common practices of Agile small teams). In McChrystal's daily briefing session, all the actors could see what was going on, grasp the connections between different actions, and make their contribution. The daily briefing that took place in the open space was a physical embodiment of the kind of collaboration that McChrystal envisaged. Electronic connections enabled units in the partner organizations—at the CIA, FBI, NSA, and conventional military units elsewhere in Iraq, the United States, and Europe—to listen in and participate. The daily give-and-take among people at all levels in the hierarchy was a dramatic embodiment of the fluid communication and collaboration needed for success.

The daily briefings were a constant affirmation that there was no established script that was being followed willy-nilly. Leaders were engaging others at the level of choice. There was no telling in advance what would be decided. The leaders did not have an outcome that was being imposed. They were open to possibility and surprise. Everything that was said was of potential consequence. The leaders had to avoid acting as though they were pressing for a predetermined conclusion. They had to be willing to allow for alternative outcomes, whatever the cost to their own egos or the pride of their own units. They had to accept that the best answer might come from anywhere in the network.

Third, McChrystal pushed *decision making and ownership* down to the lowest possible level for every action. In effect, McChrystal was creating an operation that functioned as a horizontal network based on trust, common purpose, shared consciousness, and empowered execution.

He largely removed himself from the decision-making process. Rather than having to approve all important operational decisions, he created an environment where those closest to the action could make immediate decisions. This meant only occasionally demanding the final say. McChrystal had to honor the judgment of his people, whenever he could. There were times of course when he had to "make the call." McChrystal occasionally pulled rank—because of external constraints or significant risk of failure. But he didn't do it often, or lightly. He had to recognize that "less is more."[14]

This in turn meant creating high levels of trust so that teams accepted the responsibility to make decisions without seeking approvals. Instead of holding people accountable, he had to *help people be accountable to each other*. McChrystal created public forums for network members to discuss and evaluate each other's actions; these forums also showcased members as examples to encourage higher performance.

Fourth, there was an *exchange of staff between teams*. McChrystal would "take an individual from one team—say, an Army Special Forces operator—and assign him to a different part of [the] force for six months—a team of SEALs, for example, or a group of analysts." As

McChrystal writes, "Predictably, initial resistance was intense. 'Our teams train in entirely different ways,' I was informed. I was told that I needed to understand that the tight bonds inside assault teams came from working with trusted comrades over years—to insert an outsider is an unwise and unfair risk to operators already performing the most difficult of missions. Simply put, it was anathema to the entire history and tradition of special operations selection, training, and war fighting."[15]

But McChrystal persisted. "Although it was a 'forced' initiative, once the mandate was in place, elite units were naturally incentivized to send their best operators and leaders. These individuals would be representing their organization, so unit pride would drive them to select the best examples from an already highly selective sample set. Many of these top-of-the-pack personalities were also the types that had a natural ability to connect with others—especially in an environment where leadership and one's capability as an operator were a critical measuring stick among peers."[16]

Fifth and perhaps the most difficult, McChrystal had to *unlearn what it means to be a leader.* A great deal of what he thought he knew about how the world worked and his role as a commander had to be discarded. He writes:

I began to view effective leadership in the new environment as more akin to gardening than chess. The move-by-move control that seemed natural to military operations proved less effective than nurturing the organization—its structure, processes, and culture—to enable the subordinate components to function with "smart autonomy." It wasn't total autonomy, because the efforts of every part of the team were tightly linked to a common concept for the fight, but it allowed those forces to be enabled with a constant flow of "shared consciousness" from across the force, and it freed them to execute actions in pursuit of the overall strategy as best they saw fit. Within our Task Force, as in a garden, the outcome was less dependent on the initial planting than on consistent maintenance. Watering, weeding, and protecting plants from rabbits and disease are essential for success. The gardener cannot actually "grow" tomatoes, squash, or beans— she can only foster an environment in which the plants do so.

Although I recognized its necessity, the mental transition from heroic leader to humble gardener was not a comfortable one. From that first day at West Point, I'd been trained to develop personal expectations and behaviors that reflected professional competence, decisiveness, and self-confidence. If adequately informed, I expected myself to have the right answers and deliver them to my force with assurance. Failure to do that would reflect weakness and invite doubts about my relevance. I felt intense pressure to fulfill the role of chess master for which I had spent a lifetime preparing. But the choice had been made for me. I had to adapt to the new reality and reshape myself as conditions were forcing us to reshape our force. And so I stopped playing chess, and I became a gardener.[17]

Implementing the Law of the Network led to a remarkable acceleration in the flow of information and the speed of decisions. It was not as though the Task Force had upped its game through the discovery of some secret treasure trove of data or a technological breakthrough in surveillance. What made the difference was that the Task Force was no longer a rigid machine. It had become "an adaptable, complex organism, constantly twisting, turning, and learning to overwhelm its protean adversary." It had mastered the Law of the Network.[18]

■

"Until the 1980s," writes Carlota Perez in her book *Technological Revolutions and Techno-Economic Paradigms*, "the prevalent organization was the one that served as the optimal framework for deploying the mass-production revolution: the centralized, hierarchical pyramid with functional compartments. The structure was applied by almost every corporation. . . . With the advent of computers and the Internet, large pyramids now appear rigid and clumsy. In its place the decentralized flexible network structure, with a strategic core and rapid communication, has shown the capacity for accommodating much larger and more complex global organizations as well as smaller ones."[19] Perez predicted that the network model would eventually encompass a very wide range of organizations, both public and private.

The network model of organization is key to making the whole organization Agile. As McChrystal found in a military setting, merely plugging Agile teams into a pyramidal hierarchy doesn't lead to the fluid horizontal communications necessary in a fast-changing context. Instead, communications tend to flow vertically, up and down the chain of command. The customer or end-user is often not in the picture at all. Many opportunities for improved performance are missed. In this setting, small gains by individual teams are possible, but the whole organization continues to operate as a clumsy, slow-moving bureaucracy.

To the conventional management thinker, large, efficient networks are an oxymoron. They are not managerially possible. This is the same thinking described in Chapter 2 that hinders the acceptance of the possibility of self-organizing teams that are both innovative and disciplined. Conventional managers know that a large network will be chaotic and won't be able to get things done. The human experience over thousands of years with Greek, Roman, and Chinese armies has shown that to be so. Big organizations require authority-based hierarchy. It's not even worth thinking about anything different.

Large and effective networks are also thought to be impossible because of the constraint posed by what sociologists call "the Dunbar number." This is the number that was first proposed in the 1990s by British anthropologist Robin Dunbar as the cognitive limit to the number of people with whom any individual can maintain stable social relationships—relationships in which an individual knows who each person is and how each person relates to every other person. Opinions differ as to the exact size of the limit, somewhere between 100 and 250. The commonly used value is 150.

The thinking is that in groups larger than 150, hierarchical rules and norms are required just to maintain a stable, cohesive group, let alone establish one that could implement a large, complex task. This constraint wasn't relieved by the arrival of pervasive computer communications. That's because the limit is a cognitive limit: The human brain, the sociologists said, simply can't contain the number of interrelationships involved in any larger grouping. Until the brain gets bigger, we are limited to groups no bigger than 150.

Yet over the last fifteen years, we have seen examples of large and efficient organizations operating as networks in the world of Agile, like those of McChrystal's force in Iraq. These examples include Riot Games, Spotify, the Network Management unit in Ericsson, and the Developer Division at Microsoft, all of which have more than 1,000 people collaborating on large, complex tasks in networks of Agile teams, rather than in pyramidal hierarchies.

Moreover, if we look beyond business organizations, we can see examples of even larger networks operating effectively. For instance, Alcoholics Anonymous has been operating as a network with more than 2 million members worldwide for more than seventy years, connecting more than 100,000 small groups.[20] There's no leadership or top-level manager telling the small groups what to do. The network style of governance has helped avoid many of the pitfalls that political and religious institutions have encountered.

Even more dramatic is the example of the "cellular church," a brainchild of Rick Warren, a pastor in Orange County, who created the Saddleback Church with a network of thousands of small prayer groups. It was based, writes Malcolm Gladwell, on "lots of little church cells—exclusive, tightly knit groups of six or seven who meet in one another's homes during the week to worship and pray. . . . The small group was an extraordinary vehicle of commitment. It was personal and flexible. It cost nothing. It was convenient, and every worshipper was able to find a small group that precisely matched his or her interests."[21]

What's interesting about Warren's Saddleback Church is that it's not in the self-help business of solving the individual's own problems. Like the world of Agile, its orientation is outward. It focuses attention on delivering value to others.

While we still have much to learn about the functioning of large networks, we can make some strong hypotheses as to what it takes to make them work:

1. **The network has a compelling goal.** Traditional managers are correct in thinking that a network will never work in a bureaucracy. That's because work in a bureaucracy is dominated by rules, procedures, and instructions from a boss. It's also unlikely to be effective if the goal of the firm is to make money for the shareholders and the C-suite. To function effectively, a network must be driven by a compelling shared goal. In the case of Agile management, the goal flows from the Law of the Customer: the naturally inspiring goal of adding value to customers. Once that obsession is shared throughout the organization, it matters less *who* is delivering the value. What's important is the *fact* of delivery. Like on a basketball team, it's less important who scores the goal: What matters is whether the goal is scored.

2. **The network comprises small groups.** These large networks are not built on a collection of individuals, like the old-style evangelists bringing huge crowds together in a big tent. Networks built on individuals will run into the legitimate constraint of the Dunbar number. "Successful movements," says John Hagel, the director of Deloitte's Center for the Edge, "are all organized around small, local action groups, who work together to achieve impact in very different contexts. These action groups are united by a loosely coupled network that enables them to seek help from others and to observe and learn from the diverse actions of each group what actions can achieve the greatest impact."[22] This requirement is of course consonant with the Law of the Small Team and McChrystal's "team of teams."

3. **The groups have an action orientation.** Small groups cultivate a spirit of collaboration, but it is a *particular kind* of collaboration. These groups are not about years of discussion, study, meditation, or reflection. They are about putting ideas into practice, not about abstract knowledge or even ideas for the sake of ideas themselves. These groups are about changing the world in a direction that the participants think is better in an important way.

4. **The network is the sum of the small groups.** The network doesn't *contain* the small groups; rather, it *is* the sum of the small groups. The small groups are not groups *within* the organization: Conceptually, they *are* the organization.

5. **The network's legal framework stays in the background.** In an organization functioning as a network, the legal aspects of an organization still exist. Boards of directors must be appointed. Bank accounts and contracts must have signatories. Officers are legally accountable for ensuring that money is properly spent, that compulsory government reports are signed and filed, and that any due taxes are paid. Legal procedures for recruiting, paying, and terminating any employees must be in place. But in a network, all this legal paraphernalia stays in the background: The substantive decisions are taken by the competence-based network. If that is not the case and the organization becomes a hierarchy of authority rather than a hierarchy of competence, then the network ceases to be a network: It has regressed back to a bureaucracy.

■

The organization as a network is one thing if the organization begins that way, like Riot Games, Spotify, Alcoholics Anonymous, or the "cellular church." But can a top-down bureaucracy transform itself into a network? If so, how? Three main approaches have been tried.

One possible approach is *top-down big-bang*. Here the goal is to defeat bureaucracy in one great battle—by ripping bureaucracy out and replacing it with a network. As with Mao's Cultural Revolution, the collateral damage here can be significant. People need time to adjust to any fundamental change in their organization's management model, and they need to have a stake not only implementing whatever follows bureaucracy, but in the design of it as well. And if the organization moves too fast, there's a risk of operational chaos. Moreover, a dramatic cultural change that is imposed on the organization from above may be in tension with the ethos of a self-managing organization and may create a backlash. It will generally make more sense to start as the organization means to go on. If an organization wants a highly empowered and involved workforce, then it should start by inviting everyone to help cocreate the new management model.[23] Nevertheless, the software firm Salesforce did have success with "a big bang" rollout—by carefully sidestepping the mistakes that other firms have made (see Box 4-3).

A second way is the *gradualist bottom-up approach*. The management sets out to replace bureaucracy incrementally, over a long period, working at it step-by-step. Says management guru Gary Hamel:

A lot of organizations have tried this [approach]. They have taken out a management layer or two, and simplified some particularly noisome processes. But they usually find that bureaucracy just grows back. The problem here is not a backlash, but half-hearted commitment. There's a hope that one can push back bureaucracy without challenging the fundamental assumptions on which it's built—that power trickles down; that decision rights are a function of title; that control must be imposed.

Over the decades, many organizations have run successful experiments with post-bureaucratic practices, but without a top-to-bottom commitment to bust bureaucracy, these little efforts are either soon aborted, or marginalized. No organization will ever defeat bureaucracy through half-measures—bureaucracy is too resilient. If the CEO is afraid to say "bureaucracy must die," then it won't.[24]

Even when high-performing Agile teams are plugged into a bureaucracy, there is inherent tension, as the goals and modalities are different. The situation isn't stable. Either the Agile teams will take over the organization, or more likely, the bureaucracy will crush the Agile teams.

The third approach is a *combination of top-down and bottom-up*, reflecting Carlson's Law. It's an organic process that is simultaneously revolutionary and evolutionary. In the examples of Barclays and Ericsson (discussed in Chapter 1), it was a combination of bottom-up and top-down. The Agile movement got going as a grassroots movement and then found support, encouragement, and funding from the top. When this happens, the Agile movement can have genuine enthusiasm from both the top and the bottom. Another example of this happy combination is Microsoft's Developer Division, to which we turn in the next chapter.

BOX 4-1

AGILITY THROUGH MARKET-BASED APPROACHES

One way of approaching enterprise agility on a large scale is a market-based approach. One celebrated example is Morningstar, the tomato processing firm, that has with great fanfare "abolished managers" and broken up the firm into independent profit centers.[1]

Another example is ABB, which states that its "guiding principle is to decentralize the Group into distinctive profit centers and assign individual accountability to each." ABB's business units are "self-contained and manageable units with strategic overview." ABB has organized as a federation of approximately 1,300 companies, each structured as a separate and distinct business unit. On average, each of these companies employs around 200 people in order to gain the advantages inherent in smaller organizations. Underneath these business units, monthly performance data are collected on 4,500 profit centers. These frontline companies are given complete responsibility for their balance sheet.[2]

Issues with the market-based approach to organizational structure include:

♦ If the units are truly independent profit centers, why not make them independent companies?

♦ If they are independent companies, will they be willing to collaborate on common goals?

♦ If the goal of each profit center is profit, will the unit be infected by a toxic focus on making money for the unit, rather than delivering value for the customer?

♦ Will a group of profit-making entities be able to make shifts in strategic direction when that is warranted?

Because of these issues, the market-based approach to achieving agility has its critics, one of which is Steven Sinofsky, the former president of the Windows Division at Microsoft. "Synthetic P&Ls are, always, evil," he writes. "Going back to the history of accounting, a P&L is a tool used by executives to inform decisions around resource and capital allocation, pricing, etc. In a large organization, it is very difficult to assign revenue and costs to a specific unit within a company and even more difficult to offer true span of control or accountability to a unit leader. The creation of P&Ls that attempt to represent a portion of a business inevitably lead[s] to an excess of internal focus, accountability shifting, and infighting. It is never a good idea to work with two sets of books, so unless a P&L truly represents control and accountability to a leader then it is far more likely to impede innovation, collaboration, and sharing than it is to facilitate well-informed decision making."[3]

NOTES

1. G. Hamel, "First, Let's Fire All the Managers," *Harvard Business Review*, November 2011, https://hbr.org/2011/12/first-lets-fire-all-the-managers.

2. B. Fischer, U. Lago, and F. Liu, *Reinventing Giants: How Chinese Global Competitor Haier Has Changed the Way Big Companies Transform* (San Francisco: Jossey-Bass, 2013).

3. S. Sinofsky, "Functional versus Unit Organizations," *Learning by Shipping*, December 3, 2016, https://medium.learningbyshipping .com/functional-versus-unit-organizations-6b82bfbaa57.

BOX 4-2

ACHIEVING LARGE-SCALE OPERATIONS
THROUGH PLATFORMS

One way of reaching Agility at large scale is through platforms.

For instance, through the use of a platform, Apple is able to achieve massive scale, with hundreds of thousands of individual developers devising every conceivable app to meet the infinitely variable needs of hundreds of millions of customers. As a result, each one of those hundreds of millions of customers has a product that is individually customizable to meet his or her own special needs and wants. This is something that a bureaucracy couldn't conceivably accomplish.

Another example of platforms is Autodesk, a leader in CAD/CAM software and building system modeling. "It has created a platform where companies in construction and civil engineering can draw on a growing number of apps created by an independent community. The aim is to help large construction companies simulate all aspects of a giant building project before breaking ground so they can anticipate problems and better coordinate suppliers." The ecosystem is massive, involving some 120 million designers and customers.[1]

Wikipedia shows that platforms can also work in the public sector. The biggest information product in the world—Wikipedia—is made by volunteers for free. Paul Mason in *The Guardian* argues that Wikipedia is "abolishing the encyclopedia business and depriving the advertising industry of an estimated $3 billion a year in revenue" and suggests implausibly that nonprofit cooperatives like Wikipedia could displace existing private-sector firms.[2] Wikipedia doesn't challenge the hegemony of tech companies like Google and Apple. It emulates them. Wikipedia is a public-sector version of the ecosystems at Apple and Autodesk. All function in the same manner—vast numbers of individuals pursuing a goal that they believe is worthwhile. Apple and Autodesk capture a profit from it while Wikipedia has opted not to. The model works in both private and public sector. The exemplars are complementary, not in competition.

A platform sidesteps the organizational problem of achieving agility and discipline at large scale, rather than solving it. It can be successful where there is no need for interaction, such as between Uber drivers or the homeowners of Airbnb, or where the interactions are effectively governed by agreed protocols, such as between the apps in Apple's iPhone.

But platforms have limits. A platform may not work if active interaction is needed to produce a coordinated solution, such as the design and evolution of the physical iPhone itself, which was not accomplished by a myriad of independent apps.

NOTES

1. Haydn Shaughnessy, *Shift* (Boise, ID: Tru Publishing, 2014).
2. P. Mason, "The End of Capitalism Has Begun," *The Guardian*, July 17, 2015, http://www.theguardian.com/books/2015/jul/17/postcapitalism-end-of-capitalism-begun.

BOX 4-3

"BIG BANG" CHANGE: SIX MISTAKES SALESFORCE DIDN'T MAKE

A key turning point for Salesforce occurred in 2006, when the leadership realized that innovation in the firm was starting to slow. It did what many software companies have done: It turned to Agile practices in one big-bang implementation across the entire organization—with considerable success. In doing so, Salesforce avoided six of the major mistakes that firms make in Agile implementations. What were the mistakes that Salesforce *didn't* make?

Mistake #1

Introduce the Change as Just Another Business Process

Salesforce saw that Agile involved not just the adoption of a new business process, but a fundamental transformation of the way work would be done in the company. They were introducing a new way of thinking, speaking, and acting in the workplace for both managers and workers.

The leadership at Salesforce saw that if a radically different approach to management were to be introduced in one part of the organization, there would be a tension at the interface between the part of the company still doing traditional management and the part managing work in the new way. The two parts would be operating in different modes and at different speeds. So they opted to go all-out with change across the whole organization.

Three elements helped the transition. First, the firm's on-demand software model was a natural fit for iterative methods. Second, an extensive automated test system was already in place to provide the backbone of the new methodology. And third, a majority of the R&D organization was working at the same location.

Mistake #2
Top Management Hedges Its Bets

Because Agile management represents a radical departure from the top-down bureaucracy, traditional management often adopts a wait-and-see approach. If the change succeeds, management will embrace it and celebrate it as its own. If it fails, they can say that it wasn't their idea but just another management fad that didn't work.

By contrast, the leaders at Salesforce were clear from the outset that they were committed to the change. They embraced it from the outset. They supported it as implementation proceeded. They were there at critical points in the transition, when boundaries were tested. Without strong executive support, the transition might have failed. For example, a key executive decision was to stick to the release date regardless of the content of the release. Although many teams argued for more time to complete needed features, the executive team stuck to the release date. Their ability to hold firm reinforced the principles of delivering early and often.

Mistake #3
Rigidly Apply a Methodology Conceived Elsewhere

Some firms try to implement Agile as a rigid methodology with no allowance made for the different requirements of different contexts. It's implemented with the exact terminologies, job descriptions, and procedures that have been worked out in other organizations. This can lead to considerable friction as the externally grown ideas don't exactly fit the new environment.

By contrast, Salesforce built on what had been learned in other organizations, but also adapted those lessons to its own context. A document was prepared describing the new process, its benefits, and why the firm was changing. The team held forty-five one-hour meetings with key people from all levels in the organization. Feedback from these meetings was incorporated into the document after each meeting, molding the design of the new process and creating broad organizational support for change. This open communication feedback loop allowed a large number of people to participate in the design of the change and engage actively in the solution.

One team in the organization had already successfully run a high-visibility project using iterative methods. This experience helped when the change was introduced to the other teams. Focusing on the principles rather than the mechanics also helped people understand why the firm was moving to a new way of working. When teams ran into a problem, they could refer back to the principles and adjust anything they thought did not correlate with the principles.

Mistake #4
Micromanage the Change

The Agile philosophy implies a shift in the traditional role of managers from controllers of individuals to enablers of self-organizing teams working in short iterations. It involves creating the space where those doing the work have the autonomy to apply their full talents and creativity to producing something that will delight the customer. In effect, the customer becomes the boss. For many managers and workers, these are very significant changes. If the shift itself is imposed from above in some peremptory eight-step program, there is a considerable risk that the new approach will be misinterpreted as a continuation of top-down, command-and-control management.

By contrast, the implementation of Agile at Salesforce modeled the new management philosophy of direction-setting and enablement, rather than detailed control. The change was led by a cross-functional team that was dedicated to making the change happen. This team was empowered to make decisions and used the new methodology for its own work. They brought in industry experts and other companies that had adopted similar techniques. They created a global schedule for the entire process, provided coaching and guidance, identified and removed systemic impediments to change, monitored success, and evangelized the new way of working throughout the organization.

Key features of the change included a focus on team output rather than individual productivity and cross-functional teams that met daily. All teams used a simple iterative process with a common vocabulary, with prioritized work programs for each iteration. They planned the work with user stories, estimated tasks with planning poker, and defined organizational roles using the common terminology for all teams. The result was a new release of software every thirty days.

Mistake #5
Keep Key Management Decisions Secret

Hierarchical bureaucracy is notoriously nontransparent. Each layer of the bureaucracy tends to tell the layer above and the layer below what it wants to hear, rather than everything it needs to know. CYA routines operate up and down the hierarchy. Finding out what's really going on depends on access to informal networks. When Agile is introduced into such a context, the risk of miscommunication and misunderstanding is high.

By contrast, Salesforce embraced the principle of total openness. All of the daily meetings were held in a public place so that everyone could see how things were progressing. A task board was displayed on the public lunch room wall so that everyone had access to what was going on. The willingness to share information with everyone enabled people to adapt daily to what was happening.

Mistake #6
Skimp on Training and Coaching

Studies show that the provision of external coaches have had a high payoff in terms of team productivity. Traditional management tends to ignore such studies because the mental model is that the managers are responsible for productivity. When the very future of the organization depends on success, skimping on coaching can be a highly counterproductive form of economizing.

By contrast, Salesforce put a huge emphasis on providing the needed training and coaching. The process started by sending a large group of people (initially program and functional managers) to training. Three key members from the cross-functional team developed a consolidated presentation and training deck that included concepts from the current methodology. Two-hour training sessions were held for every team. In addition, training was given to the client representatives who would be setting priorities as "product owners." They also created an internal, wiki-based website as a reference for team members as they made the transition to the new methodology and for information about the change process.

5.

IMPLEMENTING AGILE AT SCALE: MICROSOFT

The point is that the conversation takes place and it's a safe conversation to have.

—AARON BJORK

N ot long ago, Brian Harry, corporate vice president of Microsoft, received an unexpected complaint on his blog. Is Microsoft *too* focused on Agile? "It is obvious," a reader named Kasper wrote, "that you have focused on agile tools the last six months. . . . My question is how long are you going to focus on this area?"

It may come as news to some that Microsoft is focused on Agile at all. The image that many people have of this giant corporation with 2016 revenues of $85 billion and 114,000 employees is that of a giant battleship that is strong and powerful but slow to maneuver and not always customer-friendly.

In fact, that was the image that came to mind when Aaron Bjork, a group program manager in Microsoft's Developer Division, contacted me and offered to share the story of Microsoft's Agile transformation journey with the SD Learning Consortium.

The Developer Division consists of about 4,000 people, working in hundreds of teams, each team having ten to twelve people and working in three-week sprints. Bjork's specific group is Visual Studio

Team Services. This group comprises forty teams totaling about 500 people. Bjork himself looks after seven of those teams.

The Developer Division ships a range of products and services, including Visual Studio, Visual Studio Team Services, Team Foundation Server, and TypeScript. Other groups at Microsoft, like Windows, Office, and Bing, are separate, but all of them are in various stages of an Agile transformation. The Developer Division is leading the enterprise-wide journey to become more Agile. It owns the "first party engineering system charter" (1ES) and is driving that across the company. There are monthly scorecards on how the big divisions are doing in adopting it.

We ended up having several site visits, and we found that Microsoft was less like a giant battleship and more like a flotilla of speedboats operating and maneuvering in an orchestrated fashion.

■

In July 2011, when Microsoft corporate vice president Brian Harry announced a corporate commitment to Agile in a blog post that was almost a love letter to Agile, many people were skeptical. Was it conceivable that a giant firm like Microsoft could embrace Agile management?[1]

Ken Schwaber, the cofounder of the Agile methodology known as Scrum, spoke for many when in a blog post he questioned whether a big corporation like Microsoft would ever be able to emancipate itself from viewing people as "assignable, parsed, optimized resources." Just as only a very small percentage of U.S. adoptions of lean manufacturing reflect its core principle of "respecting, valuing, and engaging the workers," would the commitment to Agile announced by Microsoft be a commitment to the outward forms of Agile, but without its values? Would this be the same old "predictive manufacturing model, wrapped in Scrum tools," that would be unable to generate creative, sophisticated, quality products?[2]

Happily, we can report from our site visits that Agile values are alive and well in the Developer Division of Microsoft. The division is not only implementing Agile practices for itself, it is also promoting them for others at Microsoft. Moreover, everyone we talked to—including developers with whom we had unscripted conversations—is living, thinking, talking, and acting with Agile values. We found a pervasive Agile mindset in which respecting, valuing, and engaging those doing the work in response to customers' needs is central.

Our host for the visit, Aaron Bjork, plays golf as a hobby—an unusual choice for a software developer. He also looks after a small farm with his family in his spare time. He was a business major in college, studying finance and information systems at the University of Washington. He gave up the finance path because, as he says, he had too much energy and creativity. Finance would have bored him to death. He joined Microsoft in 2002 and started working as a coder. Seven years ago, he became what Microsoft calls a program manager, which is essentially a "product owner" in the terminology of Scrum. Why did he change? He used the analogy of a car to describe his passion. "I enjoyed having the hood up and getting my hands on the engine—getting dirty," he says. "But my real passion is deciding what kind of car we're going to build, and having input into the 'little things,' like the color of the leather stitching." Bjork knows these things matter to customers. That's his passion.

"My job," says Bjork, "is listening to what customers want and making sure that we build them something they need, and something we can sell, differentiating between wants and needs. And then finally apologizing for all the things we are not doing. There is always too much that we could be doing but can't get to. We just can't do it all."

The transformation at Microsoft took time. Aaron had begun experimenting with Agile and Scrum with his previous team back in 2008. About a year later, several teams began implementing Scrum, and there were other pockets of Agile in various places around Microsoft. In 2010, the Team Foundation Server team decided to "go Agile," with all their teams operating with Scrum practices and roles in three-week sprints, in the same cadence. Based on the success of these efforts, in July 2011, Brian Harry, the corporate vice president, publicly announced in his blog the commitment to Agile and Scrum in the Visual Studio group. In late 2011, the entire Developer Division decided to "go Agile."

By July 2015, when we first visited Microsoft, Visual Studio Team Services was in its first week of sprint #87. In May 2016, we encountered the second week of sprint #101. The three-week sprints continue in a steady rhythm, through holidays and corporate changes, come what may. Over the holidays, it's just a lighter sprint. The teams like the rhythm. They like the order that it brings.

But the Agile transformation journey has been anything but a straight path from A to B. Introducing the practices of Scrum—sprint planning, backlogs, daily standups, retrospectives—was only part of the challenge. More important—and difficult—was the shift in mindset for all involved.

It's a journey that never ends. "The journey," says Aaron, "is neither a train wreck nor a tale of unbroken triumph. It's had its ups and downs. We did some things right and some things wrong. There are places where we have improved.

"Initially there was a lot of pain. It took a long time before we could actually ship at the end of a three-week sprint," he says. "In reality, we were running three-week milestones. That's all they were. We would get to the end of a sprint and a team would claim that a feature was done and be celebrating and then I would try to use it and it wouldn't work. The team would say, 'Oh, we didn't do the setup and upgrade for it.' And I would ask, 'I thought you said it was done?' And they would reply, 'Well, yes, it's done. We just didn't do the setup and upgrade.' It took a long time for everyone to grasp that we needed to get fully done in every sprint. It took about a year to learn how to do it," he says.

"For instance, let's not be date-driven. Let's be value-driven. Let's ship the product when it's ready. We have learned not to change ship dates based on external events. If there is a problem, the team fixes it if they can. If they can't, they just don't deploy that feature. It's the team that decides. The team does its own stress-testing and documentation. It felt awkward and even weird for a long time," says Bjork. "For a lot of people, it was very disruptive. Now it's part of our DNA. It's just the way we work."

Here are the keys that Bjork believes are needed to make Agile happen at scale.

Get the Right Balance of Alignment and Autonomy

The goal at Microsoft is to have alignment at the top and autonomy at the bottom. "The teams need autonomy," says Bjork. "That's what puts a spring in their step coming to work to deliver great value to customers. But, at the same time, their work has to be aligned with the business. If there is too much control, nothing gets done: No one wants to work there and it's not a fun environment. In fact, it's a disaster. If there is too little control, it's chaos. Everyone is building whatever they want. There are no end-to-end scenarios. Customers are frustrated. Nothing makes business sense. The managers are always striving for the right balance."

The leadership team is responsible for the rules of the road. This includes clarifying the roles, the teams, the cadence, the vocabulary, and the limits on the number of quality problems—bugs—a team can have ("the bug cap").

The team has autonomy in terms of how they go about doing the work in both planning and implementation. Within the overall framework, each team can take a different approach. The specific engineering practices are up to the team. For instance, whether the team decides to do pair programming is up to them. (By comparison, at Menlo Innovations, discussed in Chapter 2, pair programming is compulsory.)

"The role of management is akin to establishing the rules of the road," says Bjork. "You can drive down the freeway fast. In fact, the freeway is designed to help you go fast. But there are rules. You must use your blinker when you cross the dotted line. You must slow down at certain points. The authorities could make the freeway much safer by putting speed bumps every hundred yards and a stop light every mile. It would be safer, but it would slow things down. At Microsoft, we take the same approach," he says. "We are specifying the minimum basic road-rules the teams need to adhere to. Our goal is to ensure that the rules help the teams move fast, to get where they want and need to go, not just slow them down."

Of course, when Bjork asks an engineering team, "What does leadership make you do that is slowing you down?" he gets a ton of stuff offloaded onto him, not just one or two things. "They often pull out a whole scroll of issues. They were just waiting for me to ask! They tell me everything I'm doing wrong and we have a conversation. The point is that the conversation takes place and it's a safe conversation to have."

Master the Role of the Agile Manager

What happens when a team fails to deliver what is expected in a sprint? A manager doesn't monitor a team's burndown chart of things-to-do. "The burndown chart is for the teams," says Bjork. "If they get behind, guess what? We talk about what to do. That's the behavior we want. We get that behavior because the culture supports it. If I yell at the team or monitor their burndown chart, guess what I get? Perfect burndown charts. I have to decide: Do I want perfect burndown charts or the right conversation? In the end, it has to be the latter."

It's the difference between mindset and practices. It's vital that people understand why they are doing the practices and then taking responsibility for the value that is meant to come with them. If the daily standups aren't working, then be a grown-up. Make a change! That's where autonomy comes in. Say Bjork, "You are in control! You are responsible!"

Handle Dependencies at the Team Level

In Microsoft's Developer Division, dependencies are handled to the extent possible by the teams themselves. The teams all know what the other teams are doing. They are all in it together. It's not like the manager or the team works only on their piece of the product and that's all they care about. They all know what other teams are doing. If one team has a dependency on another team, they are not going to wait till a meeting to let them know about it. The program managers and the engineering managers are talking about it and figuring that out ahead of time. They are self-managing and learning how to become good at it.

Of course, it sometimes happens that the team leaders get to the meeting and realize, "Oh, you guys are starting this item already? We didn't know that! You should talk to the other team." They make sure that there's a conversation and it's taken offline and they figure it out.

It's expected that the three-sprint plans include all dependencies. Bjork works with his seven teams. And the same process is happening in the six other groups of teams. The managers are talking and sorting it out on a continuous basis. They sit in the same team room. It's an ongoing conversation.

Every three months, there's a standing meeting across all of the teams. It's called "a feature team chat." Every team comes in and shares their plan. Bjork has ninety minutes. His teams come in and they each have ten to fifteen minutes to share their plan. And his peers come in with their teams and they also share. This takes place across all the teams so that everyone is aware. This regular "ceremony" provides the leadership team of the Developer Division a chance to stay in sync with what's going on.

Ensure Continuous Integration

Continuous delivery has required more modularity in design and a change in the architecture of the software. When the Developer Division first got into software as a service, they put it in the shared-computing space known as "the Cloud" and crossed their fingers. It didn't work. When one part of the product failed, the whole product failed. They saw that they needed each part of the service to be able to fail independently of every other part and not cascade through the whole product. There were deep architectural changes needed to support that.

When they started, each team would work in a branch of the code during a three-week sprint. At the end of the sprint, they would put it all together, and it was chaos. In fact, the teams had created a lot of "integration debt." That model didn't work. To ship every sprint, they had to make a fundamental change.

In principle, all the teams are "working in the same branch." What that means is that each of the teams is branching their changes in a program called Git. But the teams don't work in a silo by themselves for three weeks and then hope it all comes together. They come together all day, every day. If a team breaks the build, it fixes the build immediately. They do whatever they have to do. The longer the team waits to put code together, the greater the risk of technical and integration debt—and disaster.

The teams use what they call "feature flags." Here's a high-level description of how it works. If they are going to do something new, the very first thing that they do is to isolate the code that they are changing and build a switch into the code. It is powered by a flag in the database. It's a configuration change. When the team writes code, they write it behind the safety of the flag. At some point, when they feel it's ready, they can turn it on just for the team. That switch is not a global switch. It's a switch for an account in the system just for the team. If that goes well, then the team can turn it on for certain customers. Those customers can see it and try it. They help the team find bugs and problems. When that's done and the team is really ready, the release notes are prepared and the team announces that they are going to flip the switch for everybody. Then they go back and refactor the old code out. This enables the teams to work alongside each other on the same code without breaking one another's work.

At the end of every sprint, the team sends out an email to all 500 people in the Visual Studio Team Services group and the leadership team. They talk about what they accomplished during that sprint and what their plan is for the next sprint. And they record a three- to five-minute video. (Warning: The videos can get fancy if the teams have aspiring Hollywood directors.) The video replaces the print demo.

Keep on Top of Technical Debt

"In the old days," says Bjork, "when the code had been written, the team had a party. They were celebrating. They felt they had accomplished something. But in reality, they were sitting on a mountain of bugs. In fact, they hadn't even found all the bugs. The team then had to go back and find them all. And fix them. And they were still months away from delivering software. It was a nightmare.

"Now the number of bugs never grows," he says. "The bug cap is the number of engineers on the team times four. So, if you have ten engineers, your bug cap is forty. If you get to forty bugs, the team stops work on new features and the next sprint and gets the bug count back down below forty. It's self-managing. Teams know this. It means that we can ship product all the time because we know we're always in a healthy state."

Embrace DevOps and Continuous Delivery

In this way of working, known as DevOps, development and operations merge. The teams own the planning, execution, delivery, and operation of each new feature.

"If the service goes down," says Bjork, "then the team has to stop everything and deal with it. If bugs are found or fixes are needed, the team is responsible for fixing it. They used to have a separate support team doing that, but who wants to spend their time fixing someone else's blunders? End-to-end responsibility for quality helps keep the team committed to quality. The teams own the entire life of the feature. If the service is breaking down frequently, then it's a problem with the quality of the code. It's also a motivation to write great code. The teams are living quality on a continuing basis. They own the features they create. There's no one else to blame. And they don't have the pressure of one big release. They can stage things and resolve problems as they go.

"The change in time frame makes a huge difference. A deadline now is three weeks," says Bjork. "Three weeks is no big deal. Before, you had only two opportunities and if you missed them, you had to wait two years. Now if it's not high quality, you don't push it out. You hold it. It's disappointing that you didn't get it out. You talk about it in your retrospective. Did you do something wrong? Or did you just underestimate the level of complexity? Did you miss something? It's better to have that conversation than to have a fire drill and punish the team for not delivering what they promised, or worse, pushing out a poor-quality product.

"It's a huge change in how things happen," he adds. "Before, there was a lot of punting bugs around the organization. People pointed fingers as to who was responsible for which bug. Bugs lasted a lot longer and there were more of them. Now the team itself finds them, fixes them, and is done with it. The lead time on bugs has come way down."

Two years ago, the group flirted with the idea of daily delivery of code. But they found that customers didn't even want it. It meant too much change. There was no business need to do daily delivery. At the same time, the teams are learning to work in ever smaller chunks of work.

Continuously Monitor Progress

The teams regularly monitor how the features are being used. The results flow into the aspirational backlogs, which are called scenarios. Every month, the program manager reports out on metrics of the accounts, measuring different aspects of the service. The group is learning to become a "data-informed" business. They don't call it "data driven" because that would run the risk of missing the big picture. They use their brain and their gut feel as well as being informed by the data. The data isn't an afterthought, though. It's often the first part of the conversation.

Part of the very definition of "done" is having the right measures. The teams see this data and monitor it, both when they are testing it and as soon as it goes live. It's not something they do after they ship it. It's part of the acceptance criteria to ship.

As soon as the code ships, the team asks: How are people using it? Is it driving people through our conversion funnel? Are they becoming a dedicated account or just casual users? They use the metrics to drive the business forward.

Listen to Customer Wants, but Meet Their Needs

The program managers do all kinds of customer visits, including at the highest level at the Microsoft's Executive Briefing Center—which is all about strategy—in customer councils with folks who use the products, and regularly with the Microsoft consultants. The program managers are talking to customers on Twitter all the time.

The teams don't blindly follow what customers say. They follow what Bjork calls "the cookie principle." "If you have a plate of cookies and you ask people if they want one, they will say yes. No one turns down a cookie. If you go to a customer and ask them, 'Do you want this feature?'—guess what they say? Of course, they say yes. Why wouldn't they? It's the innovator's dilemma," he says. "There's a whole bunch of good stuff out there that you *can* do. You need to listen, but not blindly follow. The program managers need to listen to what the customers say they want, but their job is to build them something they need. And something that the firm can sell. Otherwise the managers are not doing their job."

do in the future. Like working on weekends. They can take care of their own business and not be subject to things they can't control.

Use Coaching Carefully

External coaches and trainers at Microsoft were noticeable in the site visits by their absence. There had been some coaching and basic training in Scrum at the outset. But after a while the group just started "doing it" themselves, figuring out what was working and doing more of that, and seeing what wasn't working and not doing that. Some of the Microsoft staff and managers in effect became Agile and Scrum coaches. But overall, the group itself just set out and went for it.

More recently, there's a recognition that a lot of new people have come on board who didn't do the basic training. So, thought is being given to doing more training. At the same time, there is a recognition that there is no "one size fits all" and that what works elsewhere may not fit the Microsoft culture.

Ensure Top-Level Support

Microsoft's top management was cautiously supportive about the Agile transformation at the outset. But that has changed. "There is now a broad recognition," says Bjork, "that Agile is the modern way to build software. That's not too difficult at the team level. You grab ten people and do some training and you can do it. But how do you do it across 4,000 people and stay in sync? That's the challenge. How do you do it on a large scale?"

To achieve that, the support of Corporate Vice President Brian Harry has been central. Bjork has had the benefit of working in the Developer Division where Scrum and Agile practices now have a deep foothold. "It also helped," says Bjork, "that our customers were writing user stories for us almost before we even knew what user stories were. We had to be fast learners.

"It takes time," says Bjork. "We are now seven years into this. We didn't make these changes all at once. The physical space change is one of the last changes we've made. If we had moved into a team room, put all the disciplines together, and tried to do three-week sprints, it wouldn't have worked. Agile has to evolve. Emotionally, it takes time. You can't make that change all at once."

■

An Agile transformation of an existing business is one thing. But what about creating a completely new business that creates whole new markets? It is to this issue that we now turn in Chapter 6.

BOX 5-1

FLATTENING THE HIERARCHY ISN'T THE ANSWER

One question that I ask workshop audiences is, "How many layers can an organization have and still be Agile?" I get various answers, ranging from two to seven. Of course, it's a trick question. The right answer is that it doesn't matter how many layers the organization has. What matters is the mindset.

For instance, the Microsoft Developer Division has multiple layers but it doesn't feel like a bureaucratic pyramid. Conversations are multidirectional. Anyone can talk to anyone. The right mindset creates the right spirit of conversation, curiosity, and fluidity.

By contrast, a tiny, single-layer organization can be tied up in bureaucratic knots with a single layer. It's the mindset that creates the rigidity. Decisions are based on authority, not competence, and legal boundaries spring up everywhere.

Julie Wulf, a faculty research associate of the National Bureau of Economic Research, writes:

For decades, management consultants and the popular business press have urged large firms to flatten their hierarchies. Flattening (or delayering, as it is also known) typically refers to the elimination of layers in a firm's organizational hierarchy, and the broadening of managers' spans of control. The alleged benefits of flattening flow primarily from pushing decisions downward to enhance customer and market responsiveness and to improve accountability and morale. Has flattening delivered on its promise to push decisions downward?[1]

Wulf concludes not.

She used a large-scale panel data set of reporting relationships, job descriptions, and compensation structures in a sample of over 300 large U.S. firms over roughly a fifteen-year period. This historical data analysis was complemented with exploratory interviews with executives (what CEOs say) and analysis of data on executive time use (what CEOs do). While firms have delayered, she presents evidence that flattened firms end up with more control and decision making at the top, not less. "Flattening can lead to exactly the opposite effects from what it promises to do," she writes. "In sum, flattening at the top is a complex phenomenon that in the end looks more like centralization."

Generally, when managers are talking about flattening the hierarchy, they are still thinking in hierarchical terms. They have yet to emerge from an inward-looking, pre-Copernican mindset. Changing the number of layers won't do much good until mindsets change and embrace competence-based interaction (the Law of the Network) and an external focus of delighting the customer (the Law of the Customer).

NOTE

1. J. Wulf, "The Flattened Firm," Harvard Business School, Working Paper 12-087, April 9, 2012, http://hbswk.hbs.edu/item/the-flattened-firmnot-as-advertised.

6.

FROM OPERATIONAL TO STRATEGIC AGILITY

Having a strategy suggests an ability to look up from the short term and the trivial to view the long term and the essential, to address causes rather than symptoms, to see woods rather than trees.

—**LAWRENCE FREEDMAN**[1]

As the Agile mindset and processes increasingly enter the management mainstream, firms are learning how to draw on the full talents of those doing the work, involve customers at every stage of product development, and generate innovations that customers value.

While most organizations implementing Agile management are still preoccupied with upgrading existing products and services through cost reductions, time savings, or quality enhancements for existing customers (i.e., operational Agility), they need to realize that the major financial gains from Agile management will flow from the practice of Strategic Agility—that is, generating innovations that create entirely new markets by turning noncustomers into customers. Strategic Agility is the next frontier of Agile management.

Transformation "confuses three fundamentally different categories of effort," writes Scott Anthony, managing partner of Innosight:

> The first is operational, or doing what you are currently doing, better, faster, or cheaper. Many companies that are "going digital"

fit in this category—they are using new technologies to solve old problems. . . . The next category of usage focuses on the operational model. Also called "core transformation," this involves doing what you are currently doing in a fundamentally different way. Netflix is an excellent example of this type of effort. . . . The final usage, and the one that has the most promise and peril, is strategic. This is transformation with a capital "T" because it involves changing the very essence of a company. Liquid to gas, lead to gold, Apple from computers to consumer gadgets, Google from advertising to driverless cars, Amazon.com from retail to cloud computing, Walgreens from pharmacy retailing to treating chronic illnesses, and so on.[2]

Don't get me wrong, operational Agility is a good thing. In fact, it's an increasingly necessary foundation for the survival of a firm. And it's also a precondition for achieving Strategic Agility. But in a marketplace where competitors are often quick to match improvements to existing products and services and where power in the marketplace has decisively shifted to customers, it can be difficult for firms to monetize those improvements. Amid intense competition, customers with choices and access to reliable information are frequently able to demand that quality improvements be forthcoming at no cost, or even lower cost.

Efficiency gains, time savings, and quality improvements operate within a limited frame. "The conventional view of the competitive landscape," as Clayton Christensen and his colleagues explain in their book *Competing Against Luck*, "puts tight constraints around what innovation is relevant and possible, as it emphasizes bench-marking and keeping up with the Joneses. Through this lens, opportunities to grab market share can seem finite, with most companies settling for gaining a few percentage points, within a zero-sum game."[3]

"We tend to confuse capitalism with competition," writes David Brooks, citing Peter Thiel, the creator of PayPal and author of *From Zero to One*. "We tend to think that whoever competes best comes out ahead. In the race to be more competitive, we sometimes confuse what is hard with what is valuable. The intensity of competition becomes a proxy for value. Instead of being slightly better than everybody else in a crowded and established field, it's often more valuable to create a new market and totally dominate it. The profit margins are much bigger, and the value to society is often bigger, too."[4]

This is the dark secret of the Agile management revolution: The major financial gains will come from Strategic Agility—namely, through mastering *market-creating innovation.*

Market-creating innovations are innovations that open up markets that didn't previously exist.

▶ Sometimes they transform products that are complicated, inconvenient, and expensive into things that are so much more affordable, convenient, and accessible so that many more people are able to buy and use them—for example, the personal computer.

▶ Sometimes the new products meet a need that people didn't realize they had and create a "must-have" dynamic for customers, even though the product may be relatively expensive—for example, Starbucks coffee or the iPhone.

Market-creating innovations usually don't come from resolving customer complaints or asking existing customers what they want. As Henry Ford allegedly said, if he had asked customers what they wanted they would have said a faster horse.[5] Market-creating innovations come from imagining and delivering something that delights whole new groups of customers once they realize the possibilities. Thus, no one was pressing Silicon Valley to create the personal computer, or Apple to create the iPhone, or Starbucks to create a billion new flavors of coffee. Those firms delivered something that surprised and delighted customers, once they experienced it. The product itself created the demand.

Market-creating innovations are where major revenue growth comes from. That's because they lead us to the so-called blue oceans of profitability, as W. Chan Kim and Renée Mauborgne explain in *Blue Ocean Strategy.*[6] An organization can generate high growth and profits by creating value for customers in uncontested market spaces ("blue oceans"), rather than by competing head-to-head with other suppliers in bloody shark-infested waters with known customers in existing markets ("red oceans"). By looking at the world from the customer's point of view, a firm doesn't *find* new ways to delight the customer: It *creates* them.

▶ Thus, the Cirque du Soleil was able to enter an apparently dying industry—circuses—and by eliminating animal acts, deemphasizing individual stars, and combining extreme athletic skill with sophisticated dance and music, it created the largest theatrical company in the world. While traditional circuses were going out of business, the Cirque du Soleil flourished.

▶ By looking at the world from the customer's point of view and finding ways to delight the customer, Apple has been able to take "mature" low-margin sectors—retail computers, music, mobile phones, and tablet computers—and turn them into huge money-makers. In the process, Apple, which was practically bankrupt in 1997, now has one of the world's largest market capitalizations.

In the world of Strategic Agility, we have learned that there is no such thing as a mature industry: There is only an industry to which imagination has yet to be applied.

The Principles of Strategic Agility

Strategic Agility occurs in two main ways: either as a by-product of operational Agility or as explicit initiative to generate market-creating innovation.

By-Product of Operational Agility

At Spotify, Discover Weekly was a feature intended to solve a known problem with an existing product: the difficulty that existing users were having in locating music that they would truly love, in Spotify's vast library of millions of songs. Discover Weekly not only solved that problem for existing users, but the innovation was so successful that it brought in tens of millions of new users and became in effect a brand in itself. In some countries, Discover Weekly playlists are better known than Spotify itself.

Spotify's approach to innovation is mainly based on the lean-startup principle that considers the biggest risk in innovation to be that of building the wrong thing. To reduce this risk, you start by imagining what you have in mind. Then you check whether any customer would want it. Then you build a prototype. Then you go on tweaking it, adding features that may help to monetize what you have built.

Before deciding to build a new product or feature, the teams inform themselves with research. "Do people actually want this? Does it solve a real problem for them?" Prototypes are built to give everyone a sense of what the feature might feel like and how the people might react.

Once they feel confident that the idea is promising, they go ahead and build a "minimum viable product"—just enough to fulfil the narrative, but far from being a completed product. When it is released to just a small percentage of all users, they use tools like A/B testing to monitor impact.[7] They keep adjusting the feature until they see the desired impact. Then they roll it out to the rest of the world while sorting out operational issues, like languages and scaling. In this way, Discover Weekly turned into a huge win, attracting tens of millions of new users for Spotify.

While this approach can work well in terms of improving *existing* products for *existing* users, it has several limitations in terms of systematically generating market-creating innovations.

First, in an ongoing organization as opposed to a startup, Agile teams are inevitably focused on making things better for *existing users*. If the improvement creates new markets of nonusers, that is a happy accident, not the main goal. To get more consistent success in generating market-creating innovations, *explicit attention to nonusers* is needed.

Second, market-creating innovations sometimes involve *eliminating* features, not adding or improving them. Paradoxically, less may be more. Thus, a firm may generate market-creating innovation by *eliminating* elements that it or other firms are marketing as high value for customers. The resulting simplification can sometimes perform the dual function of lowering costs and drawing in vast numbers of new users. A classic case is Southwest Airlines, which based its business on eliminating the very features the rest of the airline industry were trumpeting: meals, lounges, and seating choices.

As professors W. Chan Kim and Renée Mauborgne point out in *Blue Ocean Strategy*:

> Southwest offered high-speed transport with frequent and flexible departures at prices attractive to the mass of buyers . . . the company emphasizes only three factors: friendly service, speed, and frequent

point-to-point departures . . . it doesn't make extra investments in meals, lounges, and seating choices. By contrast, Southwest's traditional competitors invest in all the airline industry's competitive factors, making it much more difficult for them to match Southwest's prices.[8]

Southwest also made strenuous efforts to create genuine teams that had fun working together and with customers. For many customers, fewer features and lower prices, plus friendly and timely service, add up to a package that is a better deal.

Yet the decision to eliminate seemingly popular features is not one that is easily taken at the level of the Agile team. Agile teams are generally focused on responding to requests from existing customers—typically adding something that is missing. Teams are unlikely to propose or carry out experiments to eliminate key features that would bring in new customers, since it is usually assumed that even features being used by only a few existing customers must be valuable to them. Moreover, existing features typically have their own constituencies *within* the organization, so a team that has created a feature often becomes the champion for retaining and improving it. Unless there is an explicit decision at a higher level to consider eliminating features with the goal of attracting new non-customers, it is unlikely to happen. Even then, it may not work. Thus, United Airlines had difficulty emulating Southwest's low-cost model with its spin-off Ted, in part because of internal lobbies to keep running the airline the way it has always been run.

Third, market-creating innovations can lead to cannibalization of the firm's existing products and so generate a reluctance to interfere with a current revenue stream. Thus, it wasn't an easy decision within Apple to include a music-playing capability in the iPhone because it cannibalized the market for the iPod. Initially, Steve Jobs himself opposed it. But then came the realization that if Apple didn't disrupt itself, some other competitor would. Thus, Apple decided to sacrifice the revenue stream from the iPod in favor of the larger potential gains from the iPhone. Such a decision is never easy and typically it has to be taken at the highest levels of the organization.

Fourth, if corporate incentives flow to those who can show immediate results from improving an existing product for existing customers, then any slow-moving "big bets" under consideration will tend to morph into "small bets" that generate quick wins. The pressure to "get results now" will make it harder to attract top talent to work on expensive slow-gestating investments that could have huge gains. To overcome these tendencies, top management must underline the importance of winning "big bets," even if their gestation is slow, and create specific incentives to bring them to fruition.

Finally, work on improving existing features can be suitable for low-investment market creation, as at Uber and Airbnb. But it's rarely a solution for innovation that requires substantial technical innovation. Lean-startup thinking is not well suited to deal with decisions on market-creating innovations involving large investments in a new product, when no one knows in advance what the eventual product or service will look or feel like or even whether the idea will work. Often, it isn't possible initially to put a prototype in front of potential users and see whether they will use it and be thrilled by it. In most cases, there will be no "hard data" on which to base a decision. When the firm is imbued with lean-startup thinking, the firm often ends up pursuing a series of "small bets" to the neglect of "big bets."[9]

In the absence of a systematic approach to fostering market-creating innovation, decisions on such large investments will often turn on corporate politics: The loudest voice having the most hierarchical clout will end up making the call. In the absence of hard numbers, proceeding with the investment will often be perceived as presenting too great a risk and the investment will be abandoned. If a decision is finally made to go ahead with investing in a capital-intensive innovation after a bruising battle at the top, it can be hard for the organization to change course even if actual data starts to show that the firm is on the wrong track. In such situations, the firm may continue to invest in a losing proposition until it turns into a disaster that is too obvious to ignore.

Yet it doesn't have to be this way. There are well-established principles that can lead to sustained success with market-creating innovations. They involve understanding the art and science of Strategic Agility.

Market-Creating Value Propositions

A systematic approach to Strategic Agility is needed to create new markets for products or services that will enjoy strong demand and growth from both customers and noncustomers. The aim is to create products or services that enjoy little competition, precisely because they meet a need in the marketplace that is currently not being met—the so-called blue oceans of profitability.

Market-creating value propositions involve a shift in thinking from the *known* to the *unknown*—from existing products to *new products*—and from existing users to *nonusers* of the firm's products. This in turn means redefining how needs are being met and, in the process, discovering value for both customers and noncustomers from elements that lie outside current thinking, both within the firm and within the industry.

It also means a change from thinking of the *outputs* of the organization to considering *outcomes* for the customer or end-user. "Instead of thinking of your company as providing a particular type of product or service—electric power, health records management, or automobile components," writes PricewaterhouseCoopers consultant Norbert Schwieters, "think of it as a producer of outcomes. The customer needs to get somewhere, so you're not a car company; you're a facilitator of that outcome. The house is cold, so you help make it warm, possibly without supplying the necessary fuel. . . . Customers, in turn, are making fewer purchases to accumulate physical things and more purchases to achieve outcomes, convenience, and value."[10]

In considering outcomes, the firm has to be thinking more broadly than the primary function that the product or service is currently performing. Take, for instance, an airline flight. At its most basic, the airline is delivering passengers from airport A to airport B. But there are many other aspects of the customer's experience that go beyond that. It includes the experience of the customers in getting from wherever they are to airport A, and in getting from airport B to wherever they want to get to eventually. It includes the experience of getting from the front door of the airport to getting inside and sitting down in the plane, including checking in and going through security checks and baggage handling. It includes the experience of what happens on the plane in terms of the services provided, with passengers wanting comfort, convenience, and the ability to do whatever they want to do on the flight—work, rest, or be entertained. It includes the quality of the interaction with the airline staff and the other passengers. It includes services that may be taken for granted such as reliability and timeliness, as well as how problems are handled, such as weather problems, flight delays, overbooked flights, or misplaced luggage. It may include intangible deliverables, such as whether the airline or the class of service is "cool."

Thus, the outcome of even an apparently simple thing like an airline flight is in fact a complex set of many interacting elements that make up the overall experience or outcome in the customer's mind. Care needs to be taken when using language like "the job to be done," which may be taken as pointing to the primary deliverable but may miss the other kinds of customer value.

In their book *Competing Against Luck*, Christensen and colleagues argue against "defining customer needs through typical market research and then delivering against them." This can lead to a focus on "functional needs without taking into account the broader social and emotional dimensions of a customer's struggle. . . . And in many cases emotional and social could be on the same plane as functional needs—and maybe even be a driver."[11] For example, the Rolex watch is a status symbol and most of its value is based on intangible deliverables.

Outcome-oriented decision making also means a shift in thinking from looking at the industry as currently conceived. Sector boundaries as we knew them in the twentieth century are collapsing. (See Box 6-1.)

Four Components of a Market-Creating Value Proposition

A helpful playbook for developing a market-creating value proposition was pioneered by Curt Carlson and his colleagues at the Silicon Valley icon SRI International, where Carlson was president and CEO from 1999 to 2014. It is described in his book *Innovation: The Five Disciplines for Creating What Customers Want*.[12] The Innovation-for-Impact Playbook describes an organizational design and value-creation process for creating major breakthroughs, as SRI International did with HDTV, the Intuitive Surgical spin-off, and Siri, among many other developments.[13]

The playbook from SRI offers a concise but complete definition for a value proposition. There are four components—Need, Approach, Benefits per costs, and Competition—that are summarized in the mnemonic NABC. "They are the fundamentals," says Carlson. "It doesn't make sense to write up a big report until you can explain them in simple language to a knowledgeable person. Once those fundamentals are in place, the full business plan is much more efficiently developed."

Identify the Need

It begins with a focus on outcomes and the customer's need, not the firm's need for a new product or a shareholder's need for value. "Does the customer have a need for something?" writes Anand Venkataraman, a colleague of Carlson's at SRI. "How acute is this need? Would a solution be a lifesaver, a painkiller, or a supplement? Furthermore, how can you quantify the need? Is the need relevant to one person, a few people, or an entire demographic? . . . The first thing you do is to record the need as you see it and determine just how big the scope is. If it's not large enough (doesn't impact a significant number of people) can it be made to? These questions can't be answered and refined until after you've quantified the need. So as a first step, write down a tentative number on how big you think this need is."[14]

Understand noncustomers. "Obviously, the first port of call should be the customers," write the authors of *Blue Ocean Strategy*. "But you should not stop there. You should also go after noncustomers. And when the customer is not the same as the user, you need to extend your observations to the users. . . . You should not only talk to these people but also watch them in action. Identifying the array of complementary products and services that are consumed alongside your own may give insight into bundling opportunities. Finally, you need to look at how customers might find alternative ways of fulfilling the need that your product or service satisfies."[15]

Study markets. Although there is a lot of attention on global markets for products such as smartphones, most markets are fragmented in narrow segments. It can be a mistake to chase a single narrow market niche. What you need is a product or service that addresses a collection of narrower market segments. Market-creating innovation implies moving into markets that are bigger than the firm's current market.

The Approach

"It's all about how you solve this particular need of the customer," writes Venkataraman. "Here is where you'll ask yourself what your secret sauce is. It's important to have a secret sauce because that's what tells you how innovative your original idea is. Besides, a secret sauce is your entry barrier. A successful company needs an entry barrier to give it an opportunity and a kind of monopoly and incentive to develop its idea to its fullest. Without an entry barrier, rather than focus on refining the core of the idea at a time when it's needed most, you would be expending all your energy on deterring others from eating your lunch. Instead of simplifying your idea, which is the key to success, you'll end up making it more complex which spells certain doom."[16]

Think platforms. "What happens with companies that have successfully externalized," says Haydn Shaughnessy, "is that they manage to lay off a lot of the burden of change onto their ecosystem. That frees management to make decisions without having to think about all the issues of scaling the base. If you think about Apple, they were able to grow an ecosystem of somewhere in the order of 500,000 developers. This meant that management wasn't faced with the administrative burden of investing in growing an army of internal developers that made Apps. This kind of externalization relieves the burden of managing scale and enables very rapid scaling."[17]

As Schwieters writes, "The race is now on to develop and expand the platform ecosystems to deliver such outcomes for many different sectors. Amazon already provides a platform for sellers to use. Leading companies are strengthening their positions as platform providers in a wide range of industries—GE and Siemens, for example, have each developed a cloud-based system for connecting machines and devices from a variety of companies, facilitating transactions, operations and logistics and collecting and analyzing data."[18]

Acquire digital competency. Traditionally, adjacent moves were perceived as risky and firms were advised to stick to their core business. Once firms acquire competency in handling very large amounts of data and operate in a network fashion, they become able to make quite radical adjacency moves, as at Amazon. As a result, the idea of a core business itself is no longer static and fixed. Instead, a fluid core enables moves into new sectors, growing competency very quickly.

Have a bias for action. "I used to attend conferences in Europe where Nokia people would talk about the digitization of everyday life," says Shaughnessy. "It was fascinating because this was 2005. They were talking about how we would end up digitizing absolutely everything. The problem is that it was Facebook that connected people. It was Google that sold the ads on the Web, and it was Apple that made the smartphone. So for all its knowledge, Nokia didn't execute. It comes back to the business of managing adjacencies. What Nokia needed to do in 2005 was to commit to its vision of the world. It had a fabulous vision. But all it did with its vision was to carry on making phones with keyboards."[19]

Build on an existing strength. John Hagel gives the example of State Street Bank. It "started as a very conventional retail bank. It was founded in 1793. In the 1970s, it faced increasing pressure in its core business. A C-suite executive realized that they needed to find a different way of doing business and came up with the idea of renting out some of their transaction processing capabilities to other banks, who were facing similar pressure. It met a need in the marketplace and they scaled that very rapidly. Over time, they walked away from their traditional core business. It gave them a new way to define their business, their processes and operations, their approaches, and their culture. It served them well."[20]

Scale. The firm will also need the capabilities to *operate at scale*. The idea must be more than "interesting"; it must relate to a market that is sufficiently large to warrant the investment, along with the potential capability to cope with a large market. Thus, when Apple made its move into mobile phones, it had to have an enormous transaction engine capable of dealing with billions of transactions in a year.

Benefits per Costs (for Both the Customer and the Producer)

Writes Venkataraman:

> This tells you not only what improvement your solution will make in the life of the customer, but also how much of a difference. It's important to understand that almost every significant benefit can be quantified, even if only by proxy. Sometimes we may consider benefits that seem vague and think that it's impossible to quantify them, but that's only because we haven't yet got the discipline to

look at them closely enough. We give in to the euphoria of having identified a need and run with it without critically examining it. Or we give in to the fear that if we looked at it too closely it might turn out there wasn't really a need after all. Discipline gives us the courage to transcend the fear and the willpower to resist this premature euphoria.

With patience, perseverance and practice we will learn to identify things in a customer's life that are inherently valuable. It doesn't always (rarely, in fact) boil down to the number of dollars a person would save by using an invention. Often the quantification of a benefit may be in terms of intangible but yet quantifiable things—for example, increasing the number of hours of their free time that they would spend with family and friends or on their hobbies, the number of words they have to use to communicate a particular idea, the amount of effort expended (in footsteps or calories, for instance) to get to a certain place, or the number of minutes one could be continuously immersed and engaged in an entertaining or other valuable experience. It may even be some combination or collection of multiple benefits, each of which has its own quantification cell. The important thing is to get this down, and not worry about putting down something incorrect because you will get numerous chances to go back and revise it. Remember that NABC [need, approach, benefits per costs, and competition] is an iterative framework.[21]

From the producer's perspective, the financial benefits compared to costs must also be positive, even if the gains take time to materialize. A mere hope that a path to monetization will somehow show up is not enough. A potential path to monetization must be thought through from the outset.

There may be indirect ways of achieving financial benefits for the producer without compromising the user experience, by distinguishing the user and the customer (see Box 3-2). For instance, Google search is free, but huge financial benefits accrue to Google through data-focused advertising. The service provided to users feels costless and frictionless, since the monetization of benefits for Google is happening in the background.

Competitors and Alternatives

Venkataraman further writes:

> This is what others are and could be doing to address the same need
> you have identified. Many times young innovators are bound to
> think that their idea is so radical that there exists no competition.
> But that's a mistaken notion. Every idea and proposal has compe-
> tition if we look at it closely enough. I appreciate that an empty
> slate is hard to get started on, so here's at least one competitor you
> can write down for any possible idea . . . your first competitor is
> the prospect that users will simply continue to do whatever they've
> been doing in the past to address that need. The number one com-
> petitor to your invention is the alternative of not having it.
>
> The source of the difficulty that most people have in identifying
> competition is that they always think of their approach and not
> the need when trying to find competitors. It's understandable that
> the approach takes center stage, of course, because that's where
> your secret sauce is—the thing of value you bring to the table, and
> naturally the thing you feel the greatest affinity for. But to really
> understand your competition, the NABC framework teaches you
> to step back and give up being intoxicated by the coolness of your
> approach for a bit. Think of the need and try to make a list of
> everything anyone is or could be doing to address the need. Don't
> think "Who else is using the same or similar approach as mine to
> meet that need?" but think "What have people done or could do to
> meet this need?" If you came up with the idea of sticky tape as a
> way to fix notices to doors, don't only think of glue as your com-
> petitor. Think of thumbtacks, chalk, email, Facebook, and Twitter
> as competition.
>
> Once you identify and make a list of the competitive approaches
> as exhaustively as you can, your own approach will now be one of
> those in the long list. You can now start enumerating the pros and
> cons of each approach quantitatively. Does a particular competitor
> reach the same users (market) as your idea will? Does it offer the
> same benefits? Is it cheaper or more expensive to make? And so on.
> If the answer to any of these questions is unfavorable, this is your
> chance to go back and see if either the need or the approach can be
> adjusted to accommodate this shortcoming. Feel fortunate that you

found this issue now, before investing thousands, if not millions, of dollars into productizing your originally short-sighted idea.[22]

Continue to Iterate

"When you've looked at all four components of the NABC once," says Venkataraman, "you go back and revisit the Need again, repeating the whole process as many times as needed. Chances are that your original thoughts on what you believed to be the need has changed. So you revise it." Just like a scientific theory, he says, "You start with the original need, and after having gone through one cycle of analysis, you come back and augment it to account for its shortcomings. You may patch it up here and there, or make fundamental changes. But the bottom line is that as long as the customer's need is genuine . . . every failed attempt to take it down would only have made it stronger by fortifying its weak spots. So if nothing else, the NABC practice promises to at least strengthen your value proposition."[23]

For most organizations, implementing NABC propositions will be an important shift in the organizational culture. As Carlson notes, "NABC value propositions apply to every position in a company. The framework is simple and fundamental. Having every conversation in the company start with the customer's needs is transformational." To get an idea of what this involves, Chapter 7 will cover Carlson's account of his sixteen-year stint as president and CEO of SRI International and the organizational culture change that he led.

BOX 6-1

THE COLLAPSE OF SECTOR BOUNDARIES

A 2016 PwC report, "The Future of Industries: Bringing Down the Walls,"[1] documents how the boundaries among sectors are dissolving. "The pace of technological change is creating at least the prospect of a new industrial order, in which most companies no longer operate within the comfort zones of their established sectors. Already, a few companies—Apple, Amazon, and GE, among them—have boldly and successfully moved into new industries. Now just about every other company will have to do business that way."[2]

The report cites examples of whole sectors being redefined and reinvented:

- **Telecommunications.** Telecom companies used to be in the business of routing calls and data. But now, they are becoming entertainment content companies.

- **Automakers.** The future of car manufacturing will be facilitating mobility on demand. Consumers will order cars from mobility services to suit their immediate needs.

- **Electric utilities.** The staid industry of power generation is facing a future of smart infrastructure. It begins with security and temperature control, and expands to embrace a diverse range of integrated and automated services, including larger-scale energy management, monitoring of building maintenance, city resource management, transportation efficiency, and eldercare.

- **Hardware.** The Internet of Things (IoT) is transforming hardware, as firms can add sensors to their products to enable predictive maintenance and other forms of security and monitoring.

- **Health care.** Here, too, IoT is leading to the use of sensors to provide data that health professionals can use to provide early diagnostic or real-time follow-up services.

- **Personalized services.** The IoT will greatly intensify the focus on outcomes, convenience, and value. In consumer products manufacturing, for example, the IoT makes it possible to get feedback directly from consumers.

♦ **3D printing.** Digital fabrication will make it possible to build everything from airplane parts to garden ornaments.

Thus, companies in all industries need to be ready to stretch their horizons beyond their existing businesses. "This doesn't necessarily mean the borders will disappear between all industries," writes Schwieters. "But if you are a business leader, you should expect your company's sector to be transformed, probably within a decade, by the shockwave of technology change that is upon us."[3]

NOTES

1. PricewaterhouseCoopers, "The Future of Industries: Bringing Down the Walls," PwC's Future in Sight Series, 2016, http://www.pwc.com/gx/en/industries/industrial-manufacturing/publications/pwc-cips-future-of-industries.pdf.
2. N. Schwieters, "The End of Conventional Industry Sectors," *strategy + business*, January 3, 2017, https://www.strategy-business.com/blog/The-End-of-Conventional-Industry-Sectors.
3. Ibid.

BOX 6-2

THE PATH FROM OPERATIONAL AGILITY TO STRATEGIC AGILITY

The journey toward Strategic Agility involves a natural progression.

♦ A firm may start experimenting with one or more teams with operational Agility, even though the firm as a whole lacks operational Agility (see Figure 6-1).
♦ As more and more teams take on Agile management, eventually a whole unit embraces Agile (see Figure 6-2).
♦ Then the whole firm, or a large unit, may embrace Agile management with major enhancement of the capacity to make quality improvements and efficiency gains (see Figure 6-3).

 ◆ Finally, the newfound operational Agility at the enterprise level may
  then evolve toward Strategic Agility with a capability to open up new
  markets (see Figure 6-4).

The path to strategic Agility thus passes through operational Agility.
Today, many firms are still learning how to achieve operational Agility at
the team and unit level. They have yet to achieve full operational Agility
(Figure 6-3), let alone Strategic Agility (Figure 6-4).

Surveys carried out by the SD Learning Consortium show that 80 to
90 percent of Agile teams currently perceive tension between the way
the Agile teams function and the way the whole organization operates.
Unless the tension is resolved, it can lead to abrupt abandonment of an
organizational commitment to Agile altogether, with the firm declaring

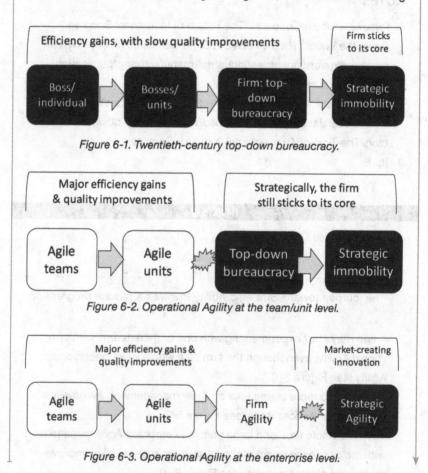

Figure 6-1. Twentieth-century top-down bureaucracy.

Figure 6-2. Operational Agility at the team/unit level.

Figure 6-3. Operational Agility at the enterprise level.

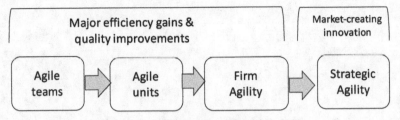

Figure 6-4. Operational and Strategic Agility.

disingenuously that "we are already Agile," only to be followed by the recognition of the sad reality shortly afterward, and a fresh effort to "implement Agile." The full gains of operational Agility only come when the whole firm embraces Agile management.

Achieving operational Agility is necessary, but it may not be sufficient to achieve financial sustainability. Even firms that achieve full operational Agility still have another frontier to master: Strategic Agility.

Operational Agility is nevertheless a crucial precondition of achieving Strategic Agility. Unless the firm is able to rapidly iterate on its market-creating initiative in accordance with the Law of the Small Team, the Law of the Customer, and the Law of the Network, there is little prospect that the market-creating initiative will succeed.

7.

CHANGING THE ORGANIZATIONAL CULTURE

Culture isn't just one aspect of the game, it is the game.

—LOUIS GERSTNER[1]

For most organizations, the shift to Agile management and Strategic Agility means a change in organizational culture—perhaps the most difficult thing that any organization can attempt. An organization's culture comprises an interlocking set of goals, roles, processes, values, communications practices, attitudes, and assumptions, many of them unspoken. The elements fit together as a mutually reinforcing system and combine to prevent any attempt to change it. That's why single-fix changes, such as the introduction of lean practices or of Agile management, may appear to make progress for a while, but eventually the interlocking elements of the organizational culture often take over and the organization is inexorably drawn back into its preexisting culture.

The bad news: Most efforts to change an organizational culture fail. This causes some people to question whether it is even worthwhile trying. Let's look at an example of success: Curt Carlson's introduction of a culture of innovation at SRI International.

In 1998, when Curt Carlson became president and CEO of SRI, it was practically bankrupt. It had been established in 1946 as a research

139

institute headquartered in Menlo Park, California. The trustees of Stanford University formed SRI as the Stanford Research Institute, a center of R&D and innovation, initially to help support economic development in California and then the world. It is now an independent company.

When Carlson became CEO, the management challenge was massive. SRI had almost run out of money. If it didn't change, it would have to sell off assets. It hadn't made a profit in years. SRI had excellent people and a legendary history as the inventor of the computer mouse, the modern personal computer interface, the first ARPA-net message, ultrasound imaging, and much more. But its organizational culture had become dysfunctional. Carlson's predecessors had tried tighter top-down control, but that crushed innovation. They had tried bottom-up innovation but that led to chaos. The organization was in a spiral of decline. Lack of trust. Lack of collaboration. Failure to share resources. Ineffective operational controls. And poor financial results. SRI's organizational culture had become so toxic that many questioned whether it could survive.

Over the next sixteen years with Carlson as CEO, SRI and its organizational culture were transformed. SRI tripled in size, became profitable, and created many world-changing innovations. One of its best-known successes was Siri, the personal assistant on the iPhone.

Under Carlson's leadership, SRI molded its organizational culture to become highly collaborative. It produced a series of market-creating innovations worth billions of dollars. I talked with Carlson about what was involved in the culture change. How did SRI become a serial innovator? Why had he succeeded where so many others have failed? What elements led to success?

Interestingly, one key element in changing SRI's culture was that Carlson never talked about "changing the culture." "People are generally proud of *their* culture," he says. "So if you go into an organization and talk about changing the culture, it makes people wonder: 'What is he talking about? What's wrong with *my* culture?' You don't want people worrying about this. I never once used the word 'culture' at SRI in any of my discussions with the staff. What I talked about was what we needed to do. I had a couple of big themes. And I repeated those themes all the time. I never used the words, 'culture change,'" he says.

What did Carlson talk about? He arrived at SRI with the notion that inventive ideas were not enough. He set out to develop a methodology for rapid, large-scale, serial innovation, starting with a focus on important customer and market needs addressed with compelling hypotheses for both the product offering and the business model. His insight was that, until that was done, all efforts at technical development would be premature, if not a total waste. He had seen most efforts at innovation fail for these reasons—no customers and no business model. His game plan included many of the hallmarks of Agile management: a focus on important customer and market opportunities; rapid, continuous team and customer cocreation; mostly self-organizing teams led by champions; positive human values and incentives; and a specific value-creation methodology that assures a high probability of success. Carlson wanted to go further and develop a value-creation playbook that would apply to the entire enterprise and generate multibillion-dollar innovations on a continuing basis.

Carlson saw that SRI had strengths on which he could build, including being in the center of Silicon Valley. The SRI staff were technically excellent, as they were in most of the companies he had worked in. The problem wasn't the staff. A key problem was the focus of the initiatives being developed. Too often, they were interesting, but not important. The initiatives were managed and developed in an ad hoc fashion. And staff members were working together mostly as individuals, not complete teams.

Carlson had the advantage of arriving at SRI with a track record of successful innovations at the Sarnoff Corporation, a subsidiary of SRI. Carlson had led the initial development of high-definition television (HDTV) and a system to assess broadcast image quality, both of which received Technology and Engineering Emmy Awards. He had pioneered the commercialization of R&D at Sarnoff and helped form over a dozen new companies while he was there.

Given this track record of accomplishment, Carlson had the support of his board, who trusted that he would succeed, even if they didn't have a deep understanding of his value-creation methods. After all, they were mainly senior operating managers—superbly accomplished professionals, but with different skills.

Carlson's track record of technical achievement also gave him credibility with the staff of SRI. He wasn't a marketing guy or a finance whiz. He had lived the life of a scientist and an engineer and had enjoyed success at the highest levels. He had written serious technical papers and given talks at professional conferences. He had the kind of credibility that is important with technical staff, who worried that an incoming CEO might be coming in to "dumb the place down."

Carlson saw that he had to change the basic way in which work got done at SRI. The previous presidents were not bad people. They were highly successful professionals in their previous roles. They were doing what they knew. The problem was that they were using an obsolete management model.

Carlson arrived with the notion that SRI had to make basic changes to focus on important customer and market needs. They would have to develop a value-creation methodology that would allow SRI to deliver higher customer value than its global competitors. The goal was to be the best at delivering new, high-value technical innovations. In a globalized economy, SRI's competition was whichever firm was the best on the planet. He also saw that if SRI was to be financially successful, it would need to be systematically hunting for "big game"—market-creating innovations with the potential of hundreds of millions of dollars of market value.

I asked Carlson what happened when he arrived at SRI and started talking like this.

"Well," says Carlson, "some of the people just loved what I was saying, and some of them hated it. They thought that aspiring to be the best in the world was ridiculous. But I was convinced we could do it, because we had achieved that at Sarnoff."

"I eventually replaced eight vice presidents," he says. "I didn't fire anybody. They were all solid professionals, but they didn't want to work in the new way and they left, one after the other. Because we were basically bankrupt when I got there, we didn't have any money to hire new people. I now think of it with some amusement as a total-immersion method for learning how to create a business!

"Even though we initially had no funds to hire new people, I didn't consider that a handicap," says Carlson. "We were able to find people who wanted to work this way. Just tremendous people. You don't need thousands of people today. You need a few excellent people who want to work collaboratively and productively on important opportunities. When they find a place to work like that, they come and they stay."

There were some difficult moments at the start. Carlson gives one example.

"In my first month, I got a phone call. I learned that a team had moved its laboratory at night from one part of the company to another without telling anybody. Imagine! I called up the vice president and said, 'Do you want to undo this?' He said, 'No, that would be too hard and destructive.' So, I called a meeting that brought everyone together," Carlson says. "I explained that from now on, we weren't going to behave that way. If anybody did this again, the entire management chain would have to go somewhere else.

"It sounds like crazy stuff, and it was exactly that: crazy stuff," he adds. "When an organization has been in decline for decades, it becomes dysfunctional. There were all kinds of team-destroying behavior going on. Every week it was something else. People often asked me why I had taken the job. It was because I knew the staff were superb and that the value-creation methodology we had pioneered at Sarnoff would be transformative. For me it was the ideal job: to be in Silicon Valley with so many brilliant colleagues."

One advantage was that Carlson had a core team of partners from the start. Without that core team, he says, it couldn't have worked. These were people who were committed to SRI's success, understood the importance of innovation, and who rolled up their sleeves to make it happen. They were working sixty-plus hours a week to make the right things materialize. He never could have made it without them.

A key person in the core team was Norman Winarsky. Carlson brought Winarsky in from SRI's subsidiary, the Sarnoff Corporation. He was one of the smartest guys Carlson had ever met. Solid. Reasonable. Fun. Great human values. And the best brainstorming partner one could imagine. At SRI, Winarsky led the new ventures and licensing practice.

Another member of the core team was Bill Wilmot, at the time a professor at the University of Montana. He was one of the world's most accomplished communication experts. He had literally written the textbook on the subject. He understood teams and behaviors, and knew how to communicate effectively.[2]

"Working with Bill," says Carlson, "was one of the best things I ever did. Every month we spent a weekend working twelve-hour days on all the people and organizational issues I was having." For example, Wilmot taught Carlson that he couldn't afford to get upset about things. You had to be firm, but not get upset. You needed to understand why people were doing those things. Many people at SRI were acting that way because they loved the place and they wanted it to be successful. But nothing was working right. It wasn't their fault. They felt helpless and afraid. They were doing what they were doing to survive.

Another key person was the head of marketing and communications, Alice Resnick. Carlson teamed with her on all communications and staff engagement activities. Carlson says, "By working closely with Alice, every communication effort was profoundly improved. We all need smart partners who add perspectives that we lack," and she was wise and perceptive about the needs of both staff and customers.

The head of human resources, Jeanie Tooker, also played a key role. She was a person with great judgment and human values. The staff completely trusted her. Later Len Polizzotto joined SRI as vice president of marketing. He had a deep understanding of the principles of value creation and added fundamental concepts. All of these people had superb human values. And they were also great fun to be with.

■

Carlson knew that he needed a systematic process for innovation. When Sarnoff became part of SRI in 1987, Carlson had worked to understand how to systematically create major new innovations and put the required organization in place. He was learning this for the first time. In big companies, value-creation skills were rarely taught and even more rarely used. "The trappings of success in a big company are beguiling," he says, "but too often completely counterproductive."

His problem was that he didn't have an efficient incubation process for creating new innovations and ventures. "We were working hard," he says, "but our initiatives didn't have the quality and customer-value needed to be successful. The biggest challenge was putting in place an organization that would make the results 'inevitable.' That is the heart and soul of the challenge and the opportunity. Getting people to change their attitude, their skills, and their ability to collaborate only comes from the way they work every day."

The thinking Carlson brought to SRI originated at Sarnoff but evolved significantly at SRI. Every other Monday night, from 5 p.m. to 9 p.m., Carlson would get his core team of fifteen people together for what he called "value-creation forums." Pizza and Coke were provided and, one after another, Carlson, Winarsky, and their colleagues would stand up and give a short value-proposition presentation for the initiative they were driving. Then the team would critique it from the perspective of what worked and what could be improved. Everyone also had to share something they had learned about innovation, markets, or potential customers. "It was a process to learn fast, not to fail fast," Carlson says. "Failing fast is a very bad idea. The goal is always to learn fast."

The process went on for about eighteen months with little success because they didn't really know what they were doing. In retrospect, Carlson says, "In those early days, we didn't understand innovation or value creation. We thought that we did. But in reality, we had been depending on the prestige and financial resources of RCA [the parent of Sarnoff]. These were super-smart people. But they didn't know how to innovate on a systematic basis. For SRI, that was going to be essential.

"For the first year, our presentations at these value-creation forums were just terrible," says Carlson. "We didn't know what a genuine value proposition looked like. What are the minimum number of questions that must be answered to have a complete value proposition? How do you know that they are good enough?"

That's how Carlson's team eventually came up with the heart of the value proposition—or what he now calls NABC (need, approach, benefits per costs, and competition)—a framework that starts the creation of any new innovation, as discussed in Chapter 6. They kept on trying different frameworks. And finally everyone said, "This is it! You've got to answer at least these four fundamental questions for any serious new innovation. You can make the list of questions as long as you want, but the list can't be smaller than this. If you take one of these four questions away, you no longer have a value proposition for innovation."

As Carlson explains, "The four questions are, What is the important customer and market need? What is your approach for addressing this need? What are the benefits per costs of your approach? And how do those benefits per costs compare with the competition and the alternatives?"

Carlson calls the answers to these four questions "a value proposition" or, in short, "NABC" for Need, Approach, Benefits per costs, and Competition. "They are the fundamentals; it doesn't make sense to write up a big report until you can explain them in simple language to a knowledgeable person."

Says Carlson, "NABC seems simple. Actually, it has proven to be profound. Because it contains the fundamental framework for creating customer value, it applies to the entire enterprise. It brings all functions together using a short, easy-to-remember meme that starts every conversation with a focus on customer needs. That is transformative. Add 'important customer and market needs' and 'constant team iteration' and you have the basic recipe for systematic organizational success. When I retired from SRI in 2014, these principles were being applied throughout the company. How many companies can say that the entire company is focused on value for their customers and that they have a memorable definition for what that means? Only a handful, in my experience."

Moreover, it wasn't enough to have a value proposition in the abstract. The value proposition had to be owned by a champion—someone who believed in the idea and would do the work necessary to make the idea a success. It meant a standard catechism of commitment, teamwork, corporate responsibility, full engagement in the value-creation process, and perseverance.

"We expected everyone to be the champion for their part," says Carlson. "We avoided the terms 'manager' and 'leader' as being confusing, inappropriate, and not collaborative enough. Champions came from all parts of the enterprise. Merit and passion were the metrics for encouragement and additional resources; not position. Our rule was, 'No champion, no project, no exception.' A champion had a family of personal attributes, human values, and high-level skills, including an understanding of what it took to innovate successfully. They were expected to be 100 percent committed to success—no excuses. We provided workshops to all staff so that they understood what this meant. We helped them gain essential value-creation skills and they would understand why, in the global innovation economy, being a champion was essential." Says Carlson, "We also made clear why it was critical for their careers to learn these skills and perform in this way."

At the same time, Carlson was doing a great deal of communication to staff. He did everything he could think of to get alignment. He gave talks. He visited the teams. He held forums where they brought people together. And every day he was at SRI, he had lunch with different staff members in the cafeteria. It didn't take long for everyone to know that if Carlson sat down with them they were going to have a discussion about their NABC value propositions and why it was so important for them to acquire SRI's value-creation skills.

Carlson and Resnick from marketing and communications started a conversation about creating an "SRI Card" that would encapsulate SRI's values and strategies. The goal of the discussion was not to produce the card. The purpose was to generate substantive discussions about SRI's needs, values, vision, and strategy. If it had taken three years to develop the consensus, that would have been fine. Until they had alignment on those basic topics, it would be hard to move the whole organization forward. So, they held meeting after meeting. They would get feedback from staff at every level. Then they would do it again. Finally, the feedback stopped coming. People started saying that they liked what they saw. That was one milestone. It took about eighteen months. Carlson wasn't in a hurry to bring it to closure. It was about having the discussion and getting agreement on how to move forward.

I asked Carlson whether there was a tipping point when people suddenly realized that things were going to be different. "There was a funny moment at the start," he replied. "Obviously, a key challenge for SRI was to start making a profit. SRI hadn't made a profit for years. So, I set a goal of making one dollar of profit for the year. We called it 'make a buck.' That may seem underwhelming, but at the time most staff thought we would fail. To make the challenge more intriguing, I promised that if we succeeded, I would play the violin at my January all-hands presentation. I had been a professional violinist at fifteen, so I thought it could work," he said. "We did a lot better than make a buck that year, but the day I played the violin was a big deal. It was a milestone. It became a public celebration that SRI was back in business and moving forward."

I asked Carlson what he did in those early years to get people's buy-in. Carlson says he included everyone but he worked mainly with the early adopters. "You never get a 100 percent," he says. "We focused on the people who wanted to work this way. You can't convert everyone on day one. That takes years.

"One essential principle in motivating professionals is to build on existing skills and values," he says. "One of the most fundamental is 'achievement.' Super-smart people will argue with you about anything and everything. But the one thing that they agree on is the desire to achieve. It's who they are. It's their identity. It's also one of the strengths of SRI—superbly smart professionals who wanted to make major, positive contributions to the world. Every talk I gave started by addressing that fundamental need—doing transformational R&D and creating world-changing innovations. Then I would describe the 'how'—our value-creation playbook."

One of the most spectacular and best-known SRI wins was Siri. Siri is an intelligent personal computer assistant and online knowledge navigator. It uses a natural-language user interface to answer questions, make recommendations, and perform actions by delegating requests to a set of web services. The software learns and adapts to the user's individual preferences and returns constantly improving individualized results. Siri was developed by SRI over a seven-year period and sold to Apple in April 2010 for hundreds of millions of dollars.

Siri illustrates the power of the NABC approach. Siri is a spin-off from SRI's Artificial Intelligence Center and an offshoot of the DARPA-funded CALO project led by SRI, which was the biggest artificial intelligence project in the history of America, with a budget of over $100 million.

"While we were doing the research on artificial intelligence for DARPA [the Defense Advanced Research Projects Agency]," says Carlson, "we were also incubating value propositions for a number of ventures. We had four or five different ideas on the table. Siri was one of them. The technical work was mostly defined but we lacked a compelling business model. So, we kept iterating value propositions without spending much money. It took two years before we finally had a good working hypothesis for the business model. Only then did we hire the best team for that value proposition. Then we iterated the business model and the product concept for another year with the new team to remove additional risks. It was only toward the end of the third year that we started building prototype products to test with early customers who were 'friends of the family.' Finally, the company, Siri Inc., was spun off in the middle of 2009.

"About six months later," says Carlson, "Steve Jobs saw Siri and he invited the Siri team to his home. We said that we didn't want to sell. But the price got to be so good that eventually the team said, 'Okay, let's sell the company to Apple. They have the platform to make it a huge innovation.'"

Siri was sold to Apple in April 2010 and it has been an integral part of the iPhone since October 2011. Siri helped Apple transform the mobile phone market and make the iPhone the dominant "must-have" mobile phone. Since then, four other companies using Siri-like technology have been spun out of SRI focused on different market opportunities.

It's easy to be dazzled by the technology involved in Siri, with its conversational interface, personal-context awareness, service delegation, and speech-recognition engine.

But Carlson stresses that the key element in the success was not the technology. It was getting the entire value proposition right. "We weren't spending a lot of money while we were working on different value propositions. One of the things that changed at SRI was the realization that we needed solid working hypotheses, both for the product and the business model, before we started spending significant money on technology. That's one of the biggest mistakes firms make," he says. "They rush ahead and want to build stuff. If we had tried to spin off the company before we had a viable business model, it would have crashed and nobody would have heard of Siri. In the 1960s to 1980s, SRI had made that mistake repeatedly. SRI spawned over twenty companies and they were mostly failures. They went out the door without a business model and they went bust.

"SRI was full of brilliant people who did amazing things," says Carlson. "But it lacked a systematic value-creation process to be fully successful. That has become even more critical for all enterprises in today's competitive environment. SRI hadn't taken advantage of all the genius it had. We had licensed the mouse to Xerox, and then to Apple, for almost nothing. SRI wouldn't do that today. That's no way to run a firm. That's also why SRI didn't get the credit from all the world-changing innovations it created."

■

Implicit in Carlson's approach to being a serial, multibillion-dollar innovator is having a long-term strategy. If a CEO is only going to be around for three years, what's their interest in doing a seven-year project?

"It's rare that you can make what Steve Jobs called 'a dent in the universe' in just three years," says Carlson. "That's a fundamental issue for many firms that chase short-term shareholder value. You don't develop multibillion-dollar ventures in a few years. And if somebody in leadership doesn't support the development of major innovations, it can be the end of the company.

"You visit these big companies. You walk in the front door and it looks like the Taj Mahal. You are expecting wonders, but you start talking to people and you find that it's just an ordinary place with dispirited staff. They aren't pursuing big ideas and, even if they are, there's no mechanism for developing them." He adds, "I often walk out of these companies depressed about the waste of the human talent working there. These companies must become profoundly more productive if they are to survive in our competitive global economy.

"The problem is that you can't learn how to launch billion-dollar innovations just by reading a book or taking a course. This is a discipline with both conceptual and experiential elements. If it's not practiced experientially," Carlson says, "the results are always poor or nonexistent. That's the big challenge we constantly see. How do you take these ideas and embed them in an organization? You need to have experience in doing that. There still aren't very many folks in senior positions with that kind of experience."

■

I asked Carlson to sum up what made the difference in changing the vision, organizational design, and spirit at SRI.

"First, you need a few really *good partners* who have the skills, values, and credibility required," he says. "People who know what to do and who have the perspectives and skills that you are missing. That is essential. You must have at least one partner; you just aren't smart enough by yourself. You need 'buddies' to test your ideas and to help you through all the challenges. The partners can be internal or external. I had both and that was ideal.

"Second," he says, "change happens in logical steps. There must first be agreement on the *need* to change. When you initially go into an enterprise suffering from challenges, people say, 'What's the need for change?' Until you have established the need, they are not ready to move.

"When people see the need, they then ask, 'What's the *vision*? Where are we going?' You are then ready to develop the vision. To gain things. Making a positive difference. That's what motivates people," he says, "particularly the kind of colleagues I had at SRI. The staff were initially worried that we were going to dumb down the place or tell them how to do their work. That's not what we had in mind—just the opposite. What we had in mind was accelerating and amplifying their achievements, both through basic research and major marketplace innovations. The only way to do that was to liberate the genius of our teams. Over time, they realized that we shared their passion and were committed to these goals.

"If I had come in and said that we are going to have to lower our standards and turn SRI into a mediocre place because that's what customers wanted, they would have thrown me out the door. But of course, many leaders make that mistake. They don't use those exact words, but basically that's the message they are delivering. In today's global innovation economy," Carlson says, "if you are not striving to be the best at what you do, you are going to disappear—fast. At the same time, we felt we were giving our colleagues many of the essential skills they needed to thrive throughout their careers. It was a huge win-win.

"You lose people if they think you're really not serious about achievement. That's what they tested me on the most. The skeptics' question was always some version of, 'Are you really serious about this, or is it just another management system *du jour*?' That was the test. As CEO, I knew that I had to walk the talk more than anyone. For example, I used the same value-creation principles as everyone else when I was the champion for a new innovation. People can immediately see what you really believe through your actions. Any lack of belief, commitment, or cynicism is deadly. I said in every way I could, 'Yes, I am serious. We are going to achieve big things. I want you and SRI to have an even bigger impact. Together we are going to change the world for the better.' I had those conversations all the time," he says.

"I constantly talked about the process for value creation, what I now call the Innovation-for-Impact Playbook.[3] It described how to work together to achieve our goals. It included the language, concepts, and processes that encapsulated how to implement the fundamentals of innovative success, such as champions, NABC value propositions, and value-creation forums. Our playbook included the importance of great human values. For example, a team will not work together with the alignment you must, as much as you can, do that in collaboration with the staff," he says.

"Third, once you have a shared vision, people start asking, 'Okay, I get the need and the vision—what's the *plan*?' It's need, vision, and plan, in that order. Many people jump to the plan before they understand the need and the shared vision. That will always get you into trouble with the staff and it is almost always wrong—as with all innovations, the first goal is to deeply understand the need," says Carlson.

"Fourth, a focus on *early adopters*. Even if people generally agree with you, most are naturally cautious. You need to find the 5 to 10 percent who will be your early champions to help you establish the principles and prove them out. Typically, 10 percent will never agree with you but once you have the early adopters on your side, you are well on your way. That's the key thing," Carlson says. "It's the champions who drive progress and become role models for everyone else. When I came to SRI, one essential person was Bill Mark, the brilliant VP of the information and computing sciences division. Bill immediately embraced the strategy, ran with it, and helped improve it. His teams created Siri and many of SRI's major innovations."

Then, "fifth, the *language* you use is critical," Carlson says. "I never used demoralizing and misleading words like 'culture change,' or 'work harder,' or 'fail fast' in my discussions with SRI staff. Rather, I talked about making a bigger impact, working smarter, and learning faster. I talked about the specific things we needed to do to be successful. To be heard and understood, you can only have two or three big actionable themes that are repeated every time. Otherwise people don't hear or understand you. Even then, getting only a few major concepts understood is not an overnight task. My themes included, one, to make an impact focus on important customer and market needs; two, to use the Innovation-for-Impact Playbook as the essential framework for success; and three, to employ intense, continuous team and customer iteration to learn fast and efficiently enough to succeed. In short: important needs, the Innovation-for-Impact Playbook, and continuous team iteration.

"I repeated these messages over and over and over. Not many themes. Just a few. Those few were always the same. Again, I would always begin by talking about achievement. Doing great things. Doing important intensity required if someone on the team is disrespectful of the others. Because collaboration is required for all major new innovations, in the global innovation economy positive human values are increasingly essential," Carlson says.

"Keeping the *support of the board* is important. In my case, they liked the success we were having even if they weren't entirely comfortable with the way we were going about it. They would be thinking, 'Presenting NABC value propositions every two weeks? Is this serious?' I never found a way to get them to really understand why these practices were fundamental to our success. The reason for this doubt is that, unless you have actually experienced the results of working this way, it can sound like a bunch of fuzzy words—or worse. To someone brought up on 400-page business plans, it doesn't seem thoughtful and structured. You need to be a value-creator yourself to appreciate that all the fundamentals of innovative success are part of the method," he says. "At its heart, it's rapid learning based on the core principles of active learning. It's very rigorous. But if you haven't experienced it, it's hard to understand at a deeper level. Yet these principles are fundamental. The next generation of innovators and management leaders will need to master them."

■

Changing a culture is thus a large-scale and long-term undertaking, involving many players. Eventually all of the organizational tools for changing minds will need to be put in play (see Figure 7-1). However, the order in which they are deployed has a critical impact on the likelihood of success. In general, the most fruitful success strategy is to begin, as Carlson did, with *leadership* tools, including a vision or story of the future, then cement the change in place with *management* tools such as role definitions and measurement and control systems, and then use the pure *power* tools of coercion and punishments as a very last resort, when all else fails.

A more draconian top-down approach to culture change took place at Apple, as sketched in Chapter 1. Steve Jobs removed thousands of top- and middle-level managers and set up his own circle of leaders to run the company. The change was successful, because the leadership was obsessed with delivering value to customers and ultimately did deliver that value and because staff believed passionately in the mission of the organization to produce insanely great products. For most organizations, the CEO will not have enough power to undertake such a change, nor the knowledge to implement it effectively. It's not a practical model to be emulated.

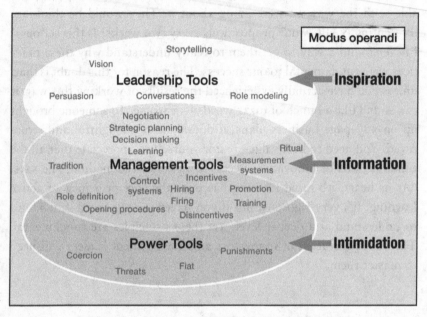

Figure 7-1. Organizational tools for changing minds.

Frequent mistakes in trying to change culture include:

- Overuse of the power tools of coercion and underuse of leadership tools
- Failure to use leadership storytelling to inspire people to embrace change[4]
- Beginning with a vision and a plan, but failing to put in place the Agile management that will cement the behavioral changes in place

BOX 7-1

SRI'S "NABC VALUE PROPOSITION" FOR SIRI

To illustrate the use of the NABC value proposition, here is a short version of the first action pitch that SRI International used for Siri, the electronic assistant that was eventually incorporated in the iPhone. "Most of what you want in a plan is still missing from a value proposition," Carlson told me. "In the case of Siri, the name 'HAL' was dropped almost immediately. But this started the conversation and then turned into a full pitch to venture capital with the main business plan questions answered. The key insight here was getting rid of the clicks.

"The key risk for us was not the technology," he said. "It was whether we could get the reference fees required for the services in the business model. Obviously, when [Steve] Jobs bought Siri, he went after the larger market as a feature on the iPhone with no reference fees. This looks obvious now, but when the team was putting this together, it wasn't."

♦ Audience: Silicon Valley Venture Capital

♦ Hook: In the movie *2001: A Space Odyssey*, HAL was a computer personal assistant. We have developed a friendly version of HAL for your mobile phone, which I would like to tell you about.

♦ Need
  • Mobile access to services is a multibillion-dollar opportunity growing at 30 percent per year.
  • Keyword search is time-consuming and ineffective, particularly on mobile devices.
  • Each "click" to find an application drops out 20 percent of users: After three to five clicks, most services and applications are effectively "lost."

♦ Approach
  • We have developed a computer personal assistant, analogous to HAL.
  • Users in English (and eventually other languages) can find and deliver information and services via speech with their mobile device.

- HAL's beachhead market is access to basic services for business travelers: Eventually it will address the much larger market for all consumers.
- The business model is reference fees from service providers.
- A prototype is developed and the first commercial product will be delivered within twelve months for $5 million.
- HAL has an outstanding team: CEO [Dag] Kittlaus, former head of Motorola's X-Team, and CTO [Adam] Cheyer, former CTO of the $100 million CALO project for DARPA.

◆ Benefits/costs
- Free app to users.
- Satisfies user's needs for basic services and completes the transactions: "Find the status of United 278," or "Get a hotel room for me tonight."
- 10X speed advantage.
- Service providers, like hotels and restaurants, get additional customers for modest charges (~$10 plus).
- A multibillion-dollar global opportunity with powerful network effects.
- HAL learns from the user, increasing accuracy over time (e.g., a preference for Marriott hotels).

◆ Competition
- World's first true computer personal assistant with a scalable business model.
- Replaces the computer mouse and keyword search.
- 2X to 10X better interface compared to mobile Google and Bing.
- Strong IP position: result of years of SRI AI research with well over $100 million invested.

◆ Close: Can I set up a meeting with you for this Friday at 9 a.m.?

[PART TWO]

MANAGEMENT
TRAPS

> *More and more corporate leaders have responded with actions that can deliver immediate returns to shareholders, such as buybacks or dividend increases, while underinvesting in innovation, skilled workforces, or essential capital expenditures necessary to sustain long-term growth.*
>
> **—LAURENCE FINK**, chief executive of BlackRock Inc.[1]

It's a familiar story. A famous global firm launches an Agile transformation, drawing on the energy and talents of bold pioneers who had been experimenting with Agile practices on their own. A C-suite executive endorses the initiative and large-scale training and coaching take place. Agile practitioners come out of the shadows and Agile management becomes part of the official life of the corporation. An Agile support team is appointed to orchestrate the initiative on a larger scale. The initiative gathers momentum as the people doing the work embrace the better way of getting things done. The faster pace of developing more value to customers is documented. Agile champions help spread the word across the organization.

The Agile initiative prospers for several years, until suddenly the C-suite declares victory. "We are Agile now," the announcement goes. "Going forward, we will be taking a more disciplined approach to delivering shareholder value."

Quiet protests ensue, but the axe comes down from above. The erstwhile sponsor in the C-suite is reassigned. The firm's Agile champions

159

are scattered or dismissed. Agile management now lacks official support in the firm and work becomes more rule-based and perfunctory. Morale suffers. Good staff members leave. Quality declines. Technical debt accumulates. Inexplicable system outages occur. Innovation grinds to a halt. After a few years, someone in the C-suite—often a newcomer—starts asking, "Why can't we get anything done? Hasn't anyone heard about Agile?" And so, the Agile transformation cycle begins again.[2]

The first part of this book (Chapters 1 through 7) dealt with the positive principles of Agile management. In this second part, Chapters 8 through 11 are dedicated to understanding why such cyclical setbacks occur and what can be done to prevent them.

The sad fact is that Agile management is at odds with much of what is practiced in public corporations and taught in business schools, even today. Generations of managers have built whole careers on different assumptions. If Agile management is to prosper in a sustainable way, leaders not only need to *learn* about Agile goals, principles, and practices. They also need to be aware of the beliefs, assumptions, and processes that they will have to *unlearn*.

Similarly, Agile leaders must recognize that they cannot proceed in a public corporation on the assumption that if teams do good work and delight their customers, they will in due course be recognized and rewarded. They need to understand what is going on elsewhere in the corporation, particularly those sections dealing with the financial aspects of management. The issues are not minor: They typically concern the very purpose of the corporation. In its crudest terms, the question is: Should the corporation be focusing on *creating* value for customers, or on *extracting* value for its shareholders and senior executives?

If Agile management is to thrive in an organization on a sustained basis, leaders need to understand why such a question has arisen, why it has become pervasive, and how it can be dealt with in a proactive fashion. In particular, corporations need to avoid four management traps:

- The trap of focusing the firm on *maximizing shareholder value* as reflected in the current stock price (Chapter 8)
- The trap of manipulating the firm's share price through *share buybacks* (Chapter 9)

- The trap of *cost-oriented economics* that focuses on short-term profits at the expense of customers and the sustainability of the corporation (Chapter 10)
- The trap of *backward-looking strategy* that deduces the future from the past, rather than using abductive logic that derives the present from a different and more fruitful future (Chapter 11)

8.

THE TRAP OF SHAREHOLDER VALUE

On the face of it, shareholder value is the dumbest idea in the world.

—JACK WELCH[1]

66 I magine an NFL coach," writes management guru Roger Martin in *Fixing the Game*, "holding a press conference on Wednesday to announce that he predicts a win by 9 points on Sunday, and that bettors should recognize that the current spread of 6 points is too low. Or picture the team's quarterback standing up in the postgame press conference and apologizing for having only won by 3 points when the final betting spread was 9 points in his team's favor. While it's laughable to imagine coaches or quarterbacks doing so, CEOs are expected to do both of these things."[2]

Imagine also, extrapolating the analogy, that the coach and his top assistants are compensated not according to whether they win games, but by whether they cover the point spread. If they beat the point spread, they receive massive bonuses. But if they miss covering the point spread a couple of times, the salary cap of the team could be cut, the coach fired, and key players released, regardless of whether the team won or lost its games.

Suppose also that, in order to manage the expectations implicit in the point spread, the coach had to spend most of his time talking with

163

analysts and sportswriters about the prospects of coming games and managing expectations about the point spread, instead of actually coaching the team. It would hardly be a surprise that the most esteemed coach would be a coach who, over multiple years, met or beat the point spread in 46 of 48 games—a 96 percent hit rate. Looking at these 48 games, one would be tempted to conclude: "Surely those scores are being fixed?"

Moreover, think what it would be like if the whole league was rife with scandals of coaches deliberately losing games ("tanking"), players sacrificing points in order not to exceed the point spread ("point shaving"), paying the referees to improve the score ("fixing the game"), or outright wagering on the results of games ("gambling").

If this were the situation in the NFL, everyone would realize that the "real game" of football had become utterly corrupted into some form of gambling and manipulation. Everyone would be calling on the NFL Commissioner to intervene and ban the coaches and players from being involved in gambling or fixing the games, and get back to playing the game of football.

Which is precisely what the NFL Commissioner did in 1962 when some players were found to be betting small sums of money on the outcome of games. In that season, Paul Hornung, the Green Bay Packers halfback and the league's most valuable player, and Alex Karras, a star defensive tackle for the Detroit Lions, were found betting on NFL games, including games in which they played. Pete Rozelle, just a few years into his thirty-year tenure as league commissioner, took immediate action. Hornung and Karras were suspended for a season. As a result, the real game of football in the NFL has remained largely separate from gambling or fixing the score. The coaches and players spend their time and effort trying to win games, not gaming the games.[3]

■

In today's paradoxical world of business, where many public corporations focus on maximizing shareholder value as reflected in the share price, the situation is the reverse. CEOs and their top managers have massive *incentives* to focus their attention on fixing the share price in the stock market, even at the cost of hurting the firm. Thus, a survey of chief financial officers has shown that "78 percent would 'give up economic value' and 55 percent would cancel a project with a positive net present value—that is, willingly harm their companies—to meet Wall Street's targets and fulfill its desire for 'smooth' earnings."[4]

The *real world of business* is the world in which factories are built, software is developed, and real products and services are designed, produced, and sold that make a real difference in customers' lives. Real dollars of earned profit show up on the bottom line. In a healthy economy, firms focus on creating real value for customers. The economy steadily grows.

The real world of business differs from the *stock market*, where shares in companies are traded between different kinds of investors. In the stock market, investors assess the resources and activities of a company today. Based on that assessment, investors form expectations as to how the company is likely to perform in future, as well as how much value the firm will return to its shareholders. The consensus view of many elements—the views of actual and potential investors, assessments of the current worth of the company, and the firm's plans for returning value to shareholders—combine to determine the stock price of the company. Some investors are in it for the long haul; others are traders out to make short-term profits from the volatility of the market.

In the current stock market, the best managers are seen as those who meet expectations. "During the heart of the Jack Welch era," writes Martin, "GE met or beat analysts' forecasts in [46 of 48] quarters between December 1989 and September 2001—a 96 percent hit rate. Even more impressively, in 41 of those 46 quarters, GE hit the analyst forecast to the exact penny—89 percent perfection. And in the remaining seven imperfect quarters, the tolerance was startlingly narrow: four times, GE beat the projection by 2 cents, once it beat it by 1 cent, once it missed by 1 cent, and once by 2 cents. Looking at these twelve years of unnatural precision . . . What is the chance that could happen if earnings were not being managed?" Martin replies: "Infinitesimal."[5]

In such a world, it takes a dedicated chief executive to do the hard, long-term work of undertaking innovation in an increasingly competitive marketplace populated with unpredictable customers. It's much simpler, easier, personally safer, and more lucrative to boost the stock price by cutting costs to enhance the apparent short-term performance in the firm's quarterly earnings, or to use financial engineering to extract value for shareholders.

In fact, a CEO of a public company has little choice but to pay careful attention to the stock market, because if the firm's share price falls markedly, accounting rules require that it be classified as a goodwill impairment.[6] Auditors may then require the company to record a loss of capital. Even more worrying, in such settings, activist hedge funds may launch campaigns to oust the management and put in place executives who are more "shareholder friendly." Thus, most executives perceive themselves to be under continuous pressure to maintain a high stock price.

By March 2016, *The Economist* could proudly declare that shareholder value thinking is now "the biggest idea in business."[7] That may be so, but it is also "the error at the heart of corporate leadership," according to two distinguished Harvard Business School professors—Joseph L. Bower and Lynn S. Paine. Maximizing shareholder value, they write, is "flawed in its assumptions, confused as a matter of law, and damaging in practice."[8]

Even Jack Welch, who was the heroic exemplar of shareholder value during his twenty-year tenure as CEO of GE, has become its harshest critic. "On the face of it," he told the *Financial Times* in 2009, "shareholder value is the dumbest idea in the world. Shareholder value is a result, not a strategy . . . your main constituencies are your employees, your customers, and your products. Managers and investors should not set share price increases as their overarching goal. . . . Short-term profits should be allied with an increase in the long-term value of a company."[9]

It often comes as a surprise to Agile practitioners that an obsession with the short-term share price should be so widespread in public corporations today. Where did this notion come from? It is often presented by its proponents as an immutable truth of the universe. Yet its origins are surprisingly recent.

∎

In the middle of the twentieth century, the conventional wisdom on running a corporation was something called "managerial capitalism." As expounded in the 1932 management classic *The Modern Corporation and Private Property* by Adolf A. Berle and Gardiner C. Means, the idea was that public firms should have professional managers who would balance the claims of different stakeholders, taking into account public policy.

This led to problems. Organizations became confused. Balancing claims by professional managers sounded good in theory. In practice, it often led to inconsistent and ill-defined priorities. Sometimes even the firms' managers couldn't understand their own processes. Decision making became unpredictable and capricious. Closure was reached only when shifting combinations of problems, solutions, and decision makers happened to coincide. Poorly understood problems wandered in and out of the system. Some theorists called it "garbage can management."[10]

That way of managing a corporation was unable to cope with the forces of globalization, deregulation, and new technology that were emerging. The careful weighing of competing stakeholder claims by professional managers was a poor fit with the faster pace of change, increased competition, and the growing power of the customer—a veritable apocalypse of change.[11]

Greater clarity in terms of purpose was needed. As discussed in earlier chapters, some firms followed Peter Drucker's lead and embraced the primacy of the *customer*. They recognized that in the emerging marketplace, firms would have to deliver more value to customers. Firms would have to become radically more innovative and nimble, just to survive. In due course, this led to the advent of management practices such as lean, design thinking, quality, and eventually Agile management.

But other public corporations headed in the opposite direction and gave primary attention to *shareholders*. Their initial champion was the Chicago economist Milton Friedman. Already, in his 1962 book, *Capitalism and Freedom*, Friedman, who was to win the Nobel Prize in Economics in 1976, had rejected Peter Drucker's view that the only valid purpose of a firm was to create a customer. Instead, he declared that "there is one and only one social responsibility of business—to use its resources and engage in activities designed to increase its profits."[12]

On September 13, 1970, Friedman set out to popularize his thinking. He published an article in *The New York Times* castigating any business managers who were not totally focused on making profits. In his view, such managers were "spending someone else's money for a general social interest."[13]

Friedman recognized that executives were not responding effectively to the apocalypse of change under way in the marketplace. His 1970 article sliced through these problems like a knife. Managers should focus on one thing and one thing only: making money for shareholders.

Everything else was irrelevant. The shareholders owned the company. Executives worked for the shareholders. The corporation was "a legal fiction." (See Box 8-1.)

The tone of Friedman's article was ferocious. Any business executives who pursued a goal other than making money for the company were "unwitting puppets of the intellectual forces that have been undermining the basis of a free society these past decades." They were guilty of "analytical looseness and lack of rigor." They had even turned themselves into "unelected government officials," who were illegally "imposing taxes" on employees, customers, and the corporation.[14]

For executives who were struggling to find their way through the ongoing apocalypse of change, Friedman's proposal offered irresistible clarity: Managers need only focus on making profits and the rest would take care of itself.

The success of Friedman's article was not because the arguments were intellectually compelling—they weren't. It's rather that executives *wanted* to believe. They were desperate to find a profitable path forward. Friedman's article was a godsend. They no longer had to worry about balancing the claims of employees, customers, the firm, and society. They could concentrate on making money for the shareholders. Adam Smith's "invisible hand" would make everything else come out right. (See Box 8-3.)

In 1976, two new champions emerged. In one of the most-cited but least-read business articles of all time, finance professors William Meckling and Michael Jensen offered a quantitative economic rationale for maximizing shareholder value, along with generous stock-based compensation to executives who followed the theory. The article explained how the personal interests of executives could be aligned with those of the corporation and its shareholders. Compensation in stock would turn the executives into part-owners of the firm and so protect the other part-owners—the shareholders—against the managers wasting cash on corporate jets, lavish new headquarters, and other monuments to executive extravagance. Now managers would act like owners. They would focus on doing "the right thing"—making money for the owners—and be well compensated for doing so.[15]

The message was seductive. It created a mandate for finance—namely, those who saw the enterprise largely through the lens of the numbers: sales figures, costs, budgets, and profits—to take charge of the corporate boardroom. Now only one thing mattered: Did it make money for the firm's owners? Since the executives themselves now were also part-owners, the approach had a happy side effect: They could become rich in the process.

In the 1980s, Ronald Reagan and Margaret Thatcher gave the idea political cover: Business should get back to basics. Government was the problem, not the solution. The business of business was the business of making money, pure and simple. Corporate raiders like Carl Icahn were happy to become the enforcers.

An important accelerator was a 1990 article in *Harvard Business Review* by finance professors Michael C. Jensen and Kevin J. Murphy. The article, "CEO Incentives—It's Not How Much You Pay, But How," suggested that many CEOs were still being paid like bureaucrats and that this caused them to act like bureaucrats. Instead, they should be paid with significant amounts of stock so that their interests would be aligned with stockholders. "Is it any wonder," Jensen and Murphy wrote, "that so many CEOs act like bureaucrats rather than the value-maximizing entrepreneurs companies need to enhance their standing in world markets?"[16]

And indeed, CEOs became very entrepreneurial—but often in *their own cause*, not necessarily the organization's cause. The article was very well received on Wall Street. The use of the phrase "maximize shareholder value" exploded. Compensation practices changed. Stock-based compensation became the norm for the C-suite. Shareholder value became the gospel of American capitalism.

Over time, corporate raiders cleverly rechristened themselves with the gentler term "activist hedge funds," as if to imply that they were performing a noble civic duty, not merely grabbing cash for themselves. But the name change didn't change the game plan.

Activist hedge funds set out to extract value even more systematically, by purchasing equity stakes in corporations and enlisting other shareholders—even public-sector pension funds—in carefully orchestrated campaigns to put pressure on the management to "unlock value" and so get their hands on the firms' assets.

The theory of maximizing shareholder value can be summarized thus:

A firm should focus on and dedicate its management and all employees to making as much money as it can for its shareholders. Such a focus, which utilizes executive stock-based compensation, will not only result in the greatest benefit to shareholders. It will also, through Adam Smith's "invisible hand," result in the optimal allocation of societal resources. The measure at any point in time of how well the firm is prosecuting the goal is the current stock price, which is the truest reflection of the value of the firm. The current stock price thus offers a metric by which the daily activities of a firm can be evaluated and the long-term value of the corporation promoted.

■

Like most bad ideas that achieve wide acceptance, shareholder value theory contains several grains of truth. One is that generating long-term value for shareholders is a good thing. If firms serve customers well and organize employees in ways that allow them to express their talents in service of customers, the company and shareholders will prosper and society will be better off.

A second is that a clear focus on outcomes is important to protect shareholders against managers wasting cash on various forms of executive extravagance—the so-called agency problem. Another partial truth is that having a single clear goal against which progress can be easily measured could be a distinct advantage in disentangling the conflicting priorities of the "garbage can management" that was prevalent in the mid-twentieth century.

Yet there was a risk. The pursuit of a single clear goal, if it was the wrong goal, could become a practical, financial, economic, social, and moral disaster.

Thus, while the idea of having business focus on solely making money for shareholders had a kind of street-corner logic to it, a sole focus on shareholders also had risks that its academic champions failed to anticipate. (See Box 8-4.) Over the ensuing decades, shareholder value theory not only failed on its own narrow terms of making money for shareholders, but it began destroying the productive capacity and dynamism of the entire economy.

Today, symptoms of the ensuing economic malfunction are everywhere apparent. A large pharmaceutical company systematically keeps buying other companies, cancels their drug research, and uses the resources to buy still more pharmaceutical companies—until the pyramid scheme abruptly collapses.[17] A major news organization turns a blind eye to sexual harassment to protect short-term profits.[18] A global auto manufacturer uses a software device to enable its diesel-driven cars to circumvent environmental regulations.[19] Firms rip off pension funds, even stuffing them with cheese, Scotch whiskey, and golf courses.[20]

Financial resources are being diverted from needed investments in innovation.[21] A brilliant study by economists from the Stern School of Business and Harvard Business School, entitled "Corporate Investment and Stock Market Listing: A Puzzle?," compares the investment patterns of public companies and privately held firms. They found that, keeping company size and industry constant, private U.S. companies invest nearly twice as much as those listed on the stock market: 7% of total assets versus just 4%. Contrary to what the academic economists had predicted in putting forward the theory of shareholder value, compensating executives with stock had made them *less* entrepreneurial for the firm, not more. As a result, the economic recovery from the Great Recession of 2008 was undermined and the ability of firms to innovate was restricted.[22]

The impact on the workforce was also significant. Manufacturing jobs were steadily sent to other countries with cheaper labor costs. In the process, public corporations, while presenting themselves as job creators, became net job destroyers. "In fact, between 1988 and 2011," write Jason Wiens and Chris Jackson of the Kauffman Foundation, "companies more than five years old destroyed more jobs than they created in all but eight of those years."[23] (See Figure 8-1.)

Shareholder value thinking has even put in question America's ability to compete in the international marketplace. "The basic narrative," says a comprehensive report from Harvard Business School, "begins in the late 1970s and the 1980s . . . firms invested less in shared resources such as pools of skilled labor, supplier networks, an educated populace, and the physical and technical infrastructure on which U.S. competitiveness ultimately depends."[24] The overall result: "Virtually all the net new jobs created over the last decade were in *local* businesses—government, health care, retailing—not exposed to international competition. That was a sign that the U.S. businesses were losing the ability to compete internationally." (See Chapter 10 and especially Box 10-2.)

The pursuit of short-term shareholder value has also led to firms allocating almost all the gains that flowed from workers' improvements in productivity to shareholders, including executives. This was a major shift from what had happened in prior decades when the economy had enjoyed steady broad-based growth. Thus, in the decades prior to 1980, compensation to workers and productivity had moved in lockstep. In the decades after 1980, shareholders were allocated almost all the gains (see Figure 8-2).[25] Such a radical reallocation of resources was not dictated by any economic force in the marketplace. It was a conscious decision by executives to reallocate almost all gains to one class of stakeholder—shareholders, including themselves—at the expense of all others. As the *Financial Times* has pointed out, these collective decisions represent "an overwhelming conflict of interest."[26]

Shareholder value thinking also led to other aberrant worker policies. Some 30 million workers in the United States are now bound by noncompete agreements that forbid them to leave their job to work for a competitor or to start their own competing business. Astonishingly, these agreements are "routinely applied to low-wage workers like warehouse employees, fast-food workers, and even dog sitters." And the number of employees covered by such agreements is increasing rapidly.[27] It is hard to see any possible justification for the proliferation of such agreements, except the predatory self-interest of the firms requiring them.

The overall result is a discouraged workforce. Only one in five workers is fully engaged in his or her work, and even fewer truly passionate about it.[28] Worse, one in seven workers is actively undermining the firm's goals.[29] In a marketplace where success depends on rapid innovation, a disengaged workforce is a significant handicap.

■

The impact of shareholder value thinking has been particularly visible in the financial sector, which has faced the challenge of making profits in a context of low interest rates and tepid economic growth. Initially, the practices adopted were not illegal, even if they were not in the best interests of customers or society. Financial firms got involved in *price gouging*, seeking unusual ways to levy hidden charges on customers, particularly customers who were vulnerable. Some started *gaming the system*, by betting against securities that they themselves created. Firms instituted practices that resemble *toll collecting*, such as high-speed trading, using their position to extract charges, simply because of their position in the system. Some got involved in *zero-sum proprietary trading in derivatives*, with dubious social benefit and significant risk to society. These activities were encouraged by recruitment and compensation practices that made such activities a permanent part of the Wall Street corporate culture.[30]

From here, it was a small step for these practices to slide into illegality. Many major banks were involved in a wide range of illegal activities, including abuses in foreclosure, money laundering for drug dealers and terrorists, assisting tax evasion, and misleading clients with worthless securities. The net fines and legal expenses of these wrongdoings since 2008 are more than $300 billion.[31]

And the consequences of shareholder value thinking roll on. In 2015, Citibank and JPMorgan Chase were among five banks who admitted committing felonies. Barclays, Royal Bank of Scotland, and UBS also pleaded guilty to criminal misconduct in a major international scam affecting the way interest rates are set around the world. The banks were fined almost $6 billion, but no senior executives were punished, let alone sent to prison. Only lower-level banking officials were dismissed. Despite committing felonies, the banks continue with business as usual.[32]

In 2015, JPMorgan Chase again admitted wrongdoing and agreed to pay $307 million in fines. JPMorgan had steered its clients into its own high-price mutual funds and hedge funds, even where lower-cost comparable funds were available, and failed to disclose this to its clients, as required by the rules of the Securities and Exchange Commission (SEC). According to the settlement, "undisclosed conflicts were pervasive."[33]

In 2016, Wells Fargo admitted creating millions of fake accounts to boost apparent sales—practices that had been going on for more than a decade, despite multiple efforts by lower-level managers to protest the wrongdoing.[34]

Against this background, it's hard to remember that, for several centuries, banks had generally been the bastions of morality. Today, insiders say that Wall Street's culture focuses on short-term gains for itself, paying scant attention to the impact of its actions on other people or society. Financial managers no longer act as stewards of the financial system.

The long tradition of finance as a principled, life-affirming, and morally worthy occupation is today less apparent than a scramble for cash.[35]

■

When activist shareholders talk of "unlocking value" from "underperforming companies," it's easy to be seduced into thinking that some worthwhile social purpose is being served and to miss the consequences for real people and the lives of communities. For instance, let's visit a company in Canton, Ohio, which has been manufacturing steel and bearings for almost a hundred years—the Timken Company. It's the kind of firm where making things still matters. Over the years, unlike other companies in the region, Timken had not moved production to other countries where labor is cheaper. Instead, Timken had made huge capital investments in Canton.

Timken, for example, "spent $225 million to expand its steel mill, allowing Timken to innovate, dominate the market for customized steel products that enjoy high margins and stay ahead of rivals in South Korea, Japan and Germany. . . . While the steel industry may not seem high-tech, research and innovation are critical—about a third of the products TimkenSteel sells now were developed only in the last five years."[36] Timken was a family-run company that also put a sizable amount of its profits into Canton in the form of good wages and donations to schools and arts. By any measure, Timken was not only a good corporate citizen. It was also well managed and highly profitable. In the ten years prior to June 2014, Timken's shares had been a good investment: It had outperformed the S&P 500—200 percent to 75 percent.

But that wasn't enough for the financial sector, which was represented in this story by two actors. One actor, Relational Investors, was a Californian hedge fund that professed to create long-term growth in publicly traded, underperforming companies that it believed were undervalued in the marketplace.[37] In fact, the modus operandi of such firms involves acquiring stakes in companies, pressuring them to make changes to "unlock value," insisting that firms load themselves up with debt, buy back their own shares, return assets to shareholders, and drive their share prices higher. Then the hedge fund cashes in its profits and moves on to the next victim, with no thought for the financially fragile state in which it leaves the firm it has attacked.

Relational's raid on Timken was a carefully orchestrated campaign, as it had carried out many times before, with accompanying infrastructure and finely tuned expertise to make the public case that Timken should be handing resources back to its shareholders. By contrast, Timken's unsuspecting management had no experience with what was about to happen and was ill-prepared to deal with the media and legal onslaught.

Relational's attack was successful. To enhance shareholder value, Timken was broken up into two companies: Timken and TimkenSteel. "All told, Relational acquired its stake at about $40 a share and sold at $70, reaping a 75 percent gain—$188 million."[38]

In the process, Relational created no jobs and made no products or services for any real people. It simply extracted money that had been created over the years by the hard work of Timken's managers and workers. Nor did Relational have to worry about what happens next to Timken, or to jobs in Canton. Within a year, Relational had sold all its shares in both Timken and TimkenSteel.[39]

Since then, the two Timken companies have not fared well in the stock market. The boost in the stock price engineered by Relational's restructuring of Timken proved ephemeral. As of early 2017, Timken had lost more than a third of its value, while TimkenSteel had lost more than two-thirds of its value, during a period when the S&P 500 was rising.[40]

The second—and more surprising—actor in this story is the California State Teachers' Retirement System, known as CalSTRS, representing teachers and their families in California—almost a million people. It is the eleventh largest public pension fund in the world. "CalSTRS owned only $16 million worth of Timken stock at the start of the fight, compared with $250 million for Relational, but it also has more than a billion dollars under management with Relational."[41] CalSTRS was the perfect front for Relational's raid. "Timken was a risky target for Relational's executives: They could be painted as Gordon Gekko types trying to make a fast buck by attacking a well-regarded, family-run company that had outperformed the stock market."[42] But having a pension fund like CalSTRS as a very public partner enabled Relational to convey the false narrative that in breaking up Timken, it was performing a virtuous public service: It was enabling the embattled schoolteachers of California to overcome the wicked plutocrats of Canton, Ohio.

It's a troubling story. But none of the actors in it are to be blamed for what happened. The Californian teachers and their pension fund were doing their best to stay solvent in a difficult low-interest-rate environment; most of the teachers probably never heard of the impact of Relational's campaign on the citizens of Canton, Ohio. Timken's managers did their best to prevent their organization from being disemboweled, but a family-run company was no match for sophisticated raiders like Relational. And Relational itself can no more be blamed for its actions than a cat can be blamed for attacking a mouse. It's in the very nature of a hedge fund to go after vulnerable companies with unsuspecting managers and then pump money out of them: That's what a hedge fund does. What is iniquitous is a system that leads a whole society to pursue short-term monetary gains at the expense of long-term investment that delivers real products and services to real customers and grows the economy.

And Timken is not an odd exception. It's just one of hundreds of such activist campaigns, which have been steadily increasing—from over 200 campaigns in 2013, to around 300 in 2015.[43] Even the largest companies aren't immune: A significant percentage of Fortune 100 and Fortune 500 companies have been targeted in the past few years.[44] The total number of activist campaigns remains high, as activists now also target small and midsize companies.[45]

■

A principal goal of shareholder value thinking was to solve the supposed problem of "agency," which is the risk that managers of corporations would act in their own interest, rather than the interests of the organization they were supposed to be managing. Ironically, shareholder value thinking has aggravated the problem of agency and turned it into a macroeconomic problem.

Even as nationwide corporate performance has been declining, executive compensation has been soaring. In the period 1978 to 2013, CEO compensation increased by an astonishing 937 percent, while the typical worker's compensation grew by a meager 10 percent.[46] The CEO-to-median-worker pay ratio of the 500 highest paid executives is now almost 1000:1 or $33 million, with 82 percent from stock-based pay.[47]

These egregious disparities are not simply the result of individual CEOs acting on their own. The problem is compounded by cronyism. Thus, in a study published in *The Accounting Review*, an astounding 62 percent of directors, who had a disclosed friendship with the CEO, said they would cut the budget for research and development in order to ensure the bonus for their friend, the CEO.[48]

The underlying rationale for shareholder value put forward by its initial academic champions was that giving stock to the CEOs would cause them to act as owners, patiently planning and working for the long-term good of the firm. What these professors failed to recognize was that, as the tenure of a CEO became increasingly short, executives would, unlike the permanent owner of a private firm, be tempted to grab what compensation they could while the going was good. Studies from the real world show that when executives are compensated with stock, they act very differently from private owners: They go straight for short-term gains for themselves—exactly what shareholder value thinking was meant to prevent.[49]

As Robin Harding in the *Financial Times* concludes, it is "time to stop thinking about corporate governance and executive pay as matters of equity and to regard them instead as a macroeconomic problem of the first rank."[50]

∎

For many years, it wasn't obvious what the cost was to the real economy of all these aberrant behaviors, as the cost was hidden by the effects of financial engineering. Fortunately, a magisterial study of all U.S. companies has been carried out by Deloitte's Center for the Edge. The conclusion? Over the period from 1965 to 2015, public corporations have been becoming steadily less productive. In the best measure of corporate performance—the rate of return on assets—U.S. firms are now performing only one-quarter as well as they were fifty years ago (see Figure 8-3).[51]

There's a deeper problem, though. Financial engineering has led to excessive growth of the financial sector, which is now roughly three times larger than it was a few decades ago.[52] Although this growth benefits those who work on Wall Street, from the point of view of the economy, it also results in a misallocation of financial and human resources.[53] People and money that could have been deployed in activities benefiting real people are instead deployed in socially unproductive activities that are no more useful than gambling in Las Vegas.

The negative impact on the economy is startlingly large. A study by the International Monetary Fund (IMF) quantifies the direct cost to U.S. economic growth of an oversize financial sector at around 2 percent of GDP per year.[54] (See Figure 8-4.) In other words, if the financial sector were the proper size, the U.S. economy would be enjoying a normal economic recovery of 3 percent to 4 percent per year instead of the dismal 1 percent to 2 percent average of recent years. That's a massive drag on the economy–over $300 billion in lost economic growth per year. In effect, the excessive financialization of the U.S. economy has become a major macroeconomic problem.

■

To return to our opening analogy, if these kinds of missteps were happening in the NFL, then everyone would realize that the "real game" of football had become utterly corrupted into gambling and manipulation. Everyone would be calling on the NFL Commissioner to intervene so that teams got back to playing the game of football.

Yet in business, when the "real game" of business has been utterly corrupted into various kinds of gambling and manipulation, business goes on as if nothing untoward is happening. Most regulators and legislators see nothing amiss.

Admittedly, some voices of alarm are now being heard: Business school professors Joseph L. Bower and Lynn S. Paine have denounced shareholder value thinking as "the error at the heart of corporate leadership."[55] Yet in the financial press, each incident of corporate missteps tends to be presented as an exception to the norm. "Bad things happen," *The Economist* admits, but don't blame shareholder value! "Outbreaks of madness in markets tend to happen because people are breaking the rules of shareholder value, not enacting them. This is true of the internet bubble of 1999–2000, the leveraged buy-out boom of 2004–08 and the banking crash. That such fiascos occur is a failure of governance and human nature, not of an idea."[56]

It's hard to look objectively at the evidence and exempt shareholder value thinking from blame by pointing to "failure of governance and human nature." It is the very essence of shareholder value thinking itself that activates the failure of governance and the defects of human nature, by institutionalizing executives' financial self-interest and legitimizing financial predators.[57] (See Box 8-2, What Is True Shareholder Value?)

As the "exceptions to the norm" give no sign of ending, society has to recognize that the exceptions have become the norm and that something must be done, particularly when the recurring patterns of behavior have disastrous financial, social, and macroeconomic consequences.

■

While the broader issues of policy and societal change will be discussed in Chapter 12, the more pressing question for Agile leaders at all levels of the organization is to figure out how to protect Agile management from the noxious consequences of shareholder value thinking.

One obvious option? Just say no! Some CEOs have already spoken out. In addition to Jack Welch's denunciation of "the world's dumbest idea":

- Vinci Group Chairman and CEO Xavier Huillard has called shareholder value thinking "totally idiotic."[58]
- Alibaba CEO Jack Ma has declared that "customers are number one; employees are number two, and shareholders are number three."[59]
- Paul Polman, CEO of Unilever, has denounced "the cult of shareholder value."[60]
- John Mackey, CEO of Whole Foods, has condemned businesses that "view their purpose as profit maximization and treat all participants in the system as means to that end."[61]
- Marc Benioff, chairman and CEO of Salesforce, has declared that this still-pervasive business theory is "wrong. The business of business isn't just about creating profits for shareholders—it's also about improving the state of the world and driving stakeholder value."[62]

Larry Fink, the CEO of BlackRock, the world's largest institutional investor, has written to all the CEOs of the S&P 500 and called on them to present long-term strategies. Companies are "under-investing in innovation, skilled work-forces or essential capital expenditures," he wrote.[63]

Sadly, though, most public corporations have yet to respond to the call. One reason is that they have succumbed to the siren call of the most appalling mechanism of finance engineering of them all: share buybacks.[64] It is to this issue that we turn in Chapter 9.

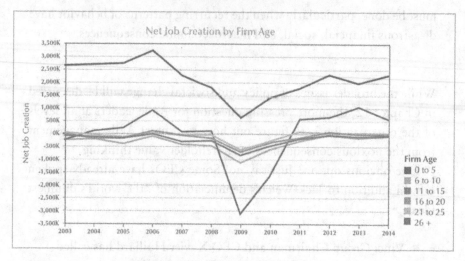

Figure 8-1. Big Firms Are Job Destroyers.

Source: J. Wiens and C. Chris Jackson, "The Importance of Young Firms for Economic Growth" The Kauffman Foundation, September 13, 2015, http://www.kauffman.org/what-we-do/resources/entrepreneurship-policy-digest/the-importance-of-young-firms-for-economic-growth.

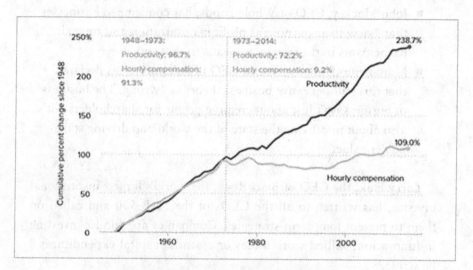

Figure 8-2. Disconnect Between Productivity
and a Typical Worker's Compensation.

Source: J. Bivens and L. Mishel, "Understanding the historic divergence between productivity and worker's pay," Economic Policy Institute, September 2, 2015, http://www.epi.org/publication/understanding-the-historic-divergence-between-productivity-and-a-typical-workers-pay-why-it-matters-and-why-its-real/.

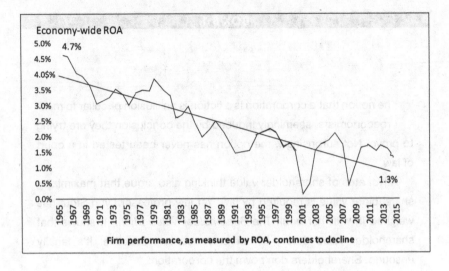

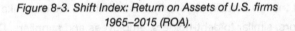

*Figure 8-3. Shift Index: Return on Assets of U.S. firms
1965–2015 (ROA).*

Source: J. Hagel, J.S. Brown, M. Wooll, and A. de Maar, "The paradox of flows: Can hope flow from fear?" Deloitte University Press, December 13, 2016, https://dupress .deloitte.com/dup-us-en/topics/strategy/shift-index.html.

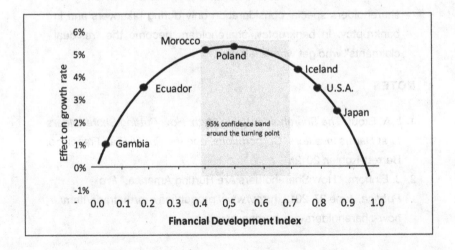

Figure 8-4: Financial Development Effect on Growth.

Source: J. Arcand, E. Berkes, and U. Panizza, "Too Much Finance," IMF Working Paper WP/12/161, June 2012, https://www.imf.org/external/pubs/ft/wp/2012/wp12161 .pdf.

BOX 8-1:

THE UNSOUND LEGAL CASE FOR
SHAREHOLDER VALUE

The notion that a corporation is a fiction is an illusion peculiar to macroeconomists, seemingly induced by the conclusion they are trying to prove. Not surprisingly, the notion has never been tested in a court of law.

Advocates of shareholder value thinking also argue that maximizing shareholder value is required by law. Yet law professor Lynn Stout has weighed in with her book *The Shareholder Value Myth*.[1] It shows that shareholder value theory is not only counterproductive: It's legally unsound. Shareholders don't own the corporation:

> What the law actually says is that shareholders are more like contractors, similar to debt-holders, employees and suppliers. Directors are not obligated to give them any and all profits, but may allocate the money in the best way they see fit. They may want to pay employees more or invest in research. Courts allow boards of directors leeway to use their own judgments. The law gives shareholders special consideration only during takeovers and in bankruptcy. In bankruptcy, shareholders become the "residual claimants" who get what's left over.[2]

NOTES

1. L.A. Stout, *The Shareholder Value Myth: How Putting Shareholders First Harms Investors, Corporations, and the Public* (San Francisco: Berrett-Kohler, 2012).
2. J. Einhorn, "How Shareholders Are Hurting America," *Pro-Publica*, June 27, 2012, http://www.propublica.org/thetrade/item/how-shareholders-are-hurting-america.

WHAT IS TRUE SHAREHOLDER VALUE?

"Today shareholder value rules business," *The Economist* proclaimed in 2016, even though, as the article admits, shareholder value thinking has "fueled a sense that Western economies are not delivering rising prosperity to most people" and is seen as "a license for bad conduct, including skimping on investment, exorbitant pay, high leverage, silly takeovers, accounting shenanigans, and a craze for share buybacks, which are running at $600 billion a year in America."[1]

"These things happen," *The Economist* admits, "but none has much to do with shareholder value." True shareholder value theory, according to the article, is something different, something pure, noble, and socially beneficent. True shareholder value is about investing in activities where "the capital employed by it made a decent return, judged by its cash-flow relative to a hurdle rate (the risk-adjusted return its providers of capital expected)."

Come again? If you didn't quite understand the last sentence, you are not alone. As the book under review in *The Economist* article (*Valuation: Measuring and Managing the Value of Companies* by McKinsey & Company) makes clear, most CEOs don't have the financial smarts to master and implement that last sentence either.

Instead, what happens, as a recent study by Credit Suisse makes clear, is that most key investment decisions are based on the reputation of the executive making the investment proposal and the CEO's "gut feel" about what should be done. Since the C-suite is hugely compensated for increases in the current stock price, guess what the "gut feel" tells the CEO? Decisions based on "shareholder value" are decisions that boost the current stock price.[2]

What rules business today is thus a degraded version of shareholder value theory—the idea that the purpose of a firm is to maximize shareholder value as reflected in the current stock price. The suggestion that most firms are basing decisions on "the capital employed by it" making "a decent return, judged by its cash-flow relative to a hurdle rate (the risk-adjusted return its providers of capital expected)" is a fantasy.

It is the short-term stock-price-oriented version of shareholder value that is presumed in daily financial news reporting, accepted as a go-to for any executive of a large public company, adopted as the *modus operandi* of activist hedge funds, endorsed by regulators, institutional investors, analysts, and politicians and often presented as simple common sense. It is this version of shareholder value that leads to the harmful consequences that *The Economist* article itself describes.

The concept of shareholder value, based on a careful calculation of "long-term shareholder value in comparison to a hurdle rate (the risk-adjusted return its providers of capital expected)" may be taught in business schools and promulgated in business textbooks like *Valuation*. But in the real world of business, it is the short-term stock-price-oriented version of shareholder value theory that is dominant.

How could it be otherwise? The very attraction of shareholder theory as articulated by Meckling and Jensen in their famous foundational article back in 1976 ("Theory of the Firm") was that it offered a clear and simple measure of performance for everyday decision making in the organization. Managers could ask themselves: "Does this action boost quarterly profits and thus enhance the stock price? If yes, do it. If not, don't." It is this kind of clarity that has led to the embrace of shareholder value theory.[3]

Sadly, it is also this kind of clarity that has led to pervasive short-term thinking, cost-cutting, crippled innovation, and all the other noxious consequences of a preoccupation with shareholder value.

NOTES

1. "Analyse This: The Enduring Power of the Biggest Idea in Business," *The Economist*, March 31, 2016, http://www.economist.com/news/business/21695940-enduring-power-biggest-idea-business-analyse.
2. Credit Suisse, "Capital Allocation—Updated," June 2, 2015, https://plus.credit-suisse.com/researchplus/ravDocView.
3. M. C. Jensen, and W.H. Meckling, "Theory of the Firm: Managerial Behavior, Agency Costs and Ownership Structure," *Journal of Financial Economics* 3, no. 4 (October 1976): 305–360, http://www.sfu.ca/~wainwrig/Econ400/jensen-meckling.pdf.

BOX 8-3

ADAM SMITH AND THE PHILOSOPHICAL ORIGINS OF SHAREHOLDER VALUE THINKING

Although shareholder value theory in its current form took shape in the 1970s, its intellectual roots go back to Adam Smith's *The Wealth of Nations* (1776), which spoke of "an invisible hand" that could miraculously turn selfishness from a vice into a virtue. A businessman might be intending his own gain, but in doing so, he was "led by an invisible hand to promote an end which was no part of his intention. . . . By pursuing his own interest, he frequently promotes that of the society more effectually than when he really intends to promote it."

It is implausible that Adam Smith himself believed in the undiluted pursuit of self-interest, since he had written an entire book suggesting the opposite: *The Theory of Moral Sentiments*.[1] Nevertheless, several centuries of economists have adopted the metaphor of "an invisible hand" and celebrated self-interest as the psychological foundation of modern economics. Elegant mathematical models—with a tenuous relationship to reality—have been developed to show that self-interest in business leads to the best of all possible worlds.

Friedman, Meckling, and Jensen thus took the preexisting metaphor of "an invisible hand" and proposed a license for enterprises to pursue unbridled self-interest across an entire society. Who can blame them for pushing the guiding metaphor of their profession to its logical conclusion? If self-interest in making money is a virtue, not a vice, why not go for it, flat out, hell for leather? Let businessmen single-mindedly pursue self-interest, and everyone will be better off!

Sadly, the theorists missed a key aspect of the fabric of reality: the Principle of Obliquity.[2] In complex social situations, objectives are often best accomplished obliquely, not directly. Central planning is not the most effective way to run an economy. Frontal assault is rarely the best military strategy. The best way to become prime minister or president isn't necessarily to declare that intent. The direct pursuit of happiness is not the best way to achieve happiness. And in business, the single-minded direct pursuit of profit isn't necessarily the best way to make a profit.

The history of business over recent decades has come to resemble a concerted effort to ignore complexity and use linear thinking to discover direct shortcuts to success. Friedman, Meckling, and Jensen embraced the embryonic psychology of mainstream economics and missed counterintuitive peculiarities of the human mind and heart. They assumed that the best way to get to a complex goal is to head for it directly. It is not. Complex settings are messy and operate in a nonlinear fashion. The actions and intentions of others, and their reactions to our actions and intentions, are key components that we have to take into account in what we plan and do. The articulation or communication of a direct goal can lead to behaviors that prevent the achievement of that goal.

Where explicit articulation of a goal will result in a complex environment pushing back in the opposite direction, an oblique goal will generally be more effective. Making money is the result, not the goal, of a successful business. It was always thus, although the shift in power in the marketplace from seller to buyer has given fresh relevance to this truth.

The purpose of a firm, and the best way to make money—indeed the only sustainable way—is to create a customer. Shortcuts to prosperity by making money out of money are dangerous illusions. Prosperity comes from generating real goods and services for real human beings, not from gambling casinos set up to generate arbitrage from volatility.

NOTES

1. *The Theory of Moral Sentiments* begins with the following assertion: "How selfish soever man may be supposed, there are evidently some principles in his nature, which interest him in the fortunes of others, and render their happiness necessary to him, though he derives nothing from it, except the pleasure of seeing it." By contrast, economists have taken the phrase in *The Wealth of Nations* about "an invisible hand" as the first step in the cumulative effort to build a model that proves that a competitive economy based on self-interest provides the largest possible economic pie.

2. J. Kay, *Obliquity: Why Our Goals Are Best Achieved Indirectly* (New York: Penguin, 2011).

BOX 8-4

THE UNANTICIPATED RISKS OF
SHAREHOLDER VALUE

The 1976 article by finance professors Meckling and Jensen is one of the most-cited but least-read business articles of all time. Replete with elaborate mathematics, it is often taken to provide a quantitative economic rationale for maximizing shareholder value, along with generous stock-based compensation to executives who followed the theory. The article, however, failed to deal with certain unanticipated risks to its central thesis.

The Meckling/Jensen article focused only on single decisions and the short-term impact of compensating executives with stock. The long-term impacts of pursuing shareholder value as reflected in the stock price were described as "important issues which are left for future analysis"—an analysis that the authors never got around to doing.

The article did envisage a risk that an executive might make "decisions which benefit him at the (short–run) expense of the current bondholders and stockholders." The article was satisfied that this risk wouldn't materialize: "If [the executive] develops a reputation for such dealings, he can expect this to unfavorably influence the terms at which he can obtain future capital from outside sources. This will tend to increase the benefits associated with 'sainthood' and will tend to reduce the size of the agency costs."

However, the article didn't foresee the risk that the "sainthood" involved in looking after the interests of shareholders might turn into ungodly combines of executives and shareholders *against* the interests of the corporation. Executives might conspire with shareholders and corporate raiders to extract value from the corporation at the expense of customers, employees, and the organization itself. Nor did the article perceive that this risk might materialize on a gargantuan scale.[1]

The theorists did not foresee the risk that if firms started to act on the basis that executives needed to be motivated to work on behalf of the shareholders for monetary gain, this might turn into a nightmarish self-fulfilling prophecy.[2] Executives might, far from aspiring to the "sainthood" of looking after the long-term interests of shareholders, turn into grotesque caricatures of self-interest and greed, grabbing extraordinary compensation and pension benefits for themselves.[3]

The theorists didn't worry about the risk to the organization itself, because to them an organization was a mere "legal fiction" (see Box 8-1). In the theoretical world inhabited by these academic economists, if one organization failed, it could be replaced by another. They didn't perceive the risk that there might be very heavy economic and social costs in dismantling corporations and even whole sectors, along with the social capital embodied in them. Nor did they perceive the risk that shifting production to other parts of the world in search of lower labor costs in order to boost shareholder returns might cause irreparable damage to the national productive capacity, or that once lost, this productive capacity might not be easily retrieved.[4]

The theorists were even less worried about risks to employees, who were regarded as assets that were fungible, and even disposable, to serve the interests of capital: If employees lost their jobs, they could always seek employment elsewhere. "The employee is protected from coercion by the employer," Milton Friedman complacently imagined, "because of other employers for whom he can work."[5] The theorists never imagined that there would come a time when tens of millions of U.S. employees—even low-level employees—would be bound by non-compete agreements that forbade employees to leave their job to work for a competitor or to start their own competing business.[6]

Nor did the theorists foresee the risk that the extraction of value from corporations might become so great that there would be few "other employers" offering quality employment. Nor did they envisage that if this process was continued over decades by most publicly owned corporations, the impacts would become macroeconomic in scale, and that there would eventually be a shortage of customers for real goods and services and limited opportunities to invest in. In the language of the macroeconomists, there would be a "lack of demand," according to Friedman. The economy could thus be facing secular economic stagnation.[7]

Nor did the theorists worry about the possibility that there would be an unholy alliance between shareholder value theory and top-down command-and-control management. Once a firm embraced maximizing shareholder value and the current stock price as its goal, and lavishly compensated top management to that end, the C-suite would have little choice but to deploy command-and-control management, because making money for shareholders and the C-suite is inherently uninspiring to employees. The C-suite would have to compel employees to obey, even if this meant that employees would become dispirited. The authors didn't worry that dispirited employees might become a critical constraint in an economy that would depend on innovation from engaged knowledge workers. How firms were managed was not a matter of much interest to academic economists.

Nor did these economic theorists recognize the risk that if corporations didn't invest in training and retraining their employees and merely shifted work to wherever labor costs were currently lowest in order to boost immediate returns to shareholders, there might come a day when there was an inadequate pool of trained workers from which they would draw on for the creation of future businesses.[8]

Nor did the theorists worry too much about customers. If customers didn't like what corporations were offering they could, as Milton Friedman had written, take their business elsewhere: "The consumer is protected from coercion by the seller because of the presence of other sellers with whom he can deal."[9]

The theorists didn't perceive the risk that if most corporations in a country favored shifting resources to shareholders instead of investing in innovation for customers, customers might become dissatisfied and take their business "elsewhere" to corporations who did care about them, so that eventually the dynamism of entire industries or even a whole national economy might be compromised.[10]

Nor did they perceive the risk of damage to communities by closing plants in order to benefit from the lower cost of labor elsewhere in the world and thereby boost returns to shareholders. They assumed that "the magic of the marketplace" and the process of "creative destruction" would heal any damage that a short-term extraction of value might cause. They didn't envisage the risk that the extraction of value might reach such a scale that there would be insufficient resources available to the economy as a whole to heal the damage to communities caused by the short-term extraction of value.[11]

Nor did they foresee the risk that diverting resources to shareholders and executives from investments in the future might reach such a scale that the capacity of the economy to grow and compete in the international marketplace or provide quality livelihood for all its citizens might be compromised.[12] They did not anticipate the possibility that the economy would ever be in a situation where it would only be able to sustain an appearance of prosperity through vast infusions of cheap money from the central bank.

Nor did they foresee the risk that by creating strong financial incentives to do the easy thing of low value to the real economy (i.e., increasing the stock price) while creating structural disincentives to do the difficult thing of high value to the organization and the real economy (i.e., grow the business by investing in market-creating opportunities), most executives would spend their efforts in self-interested activities of low value to the organization and the real economy.

Yet these were the very risks that materialized over the next four decades.

Friedman was famous for pointing to the fallacy of good intentions. "Concentrated power," he wrote in *Capitalism and Freedom*, "is not rendered harmless by the good intentions of those who create it." Friedman was referring, of course, to government, but his dictum has also turned out to be a prophetic critique of his own legacy in the private sector.

In his celebrated career, Friedman himself had great influence on economic policy and management practice. But with great influence comes great responsibility. Friedman's doctrines have had consequences that are the opposite of what he intended. Friedman was an honorable man. His intentions were good, but the consequences of what he wrote are not rendered harmless by his good intentions. His admirers claim that he "rescued the U.S. economy" and provided "a cure for capitalism."[13] Unfortunately, we can see with the wisdom of hindsight that the cure has turned out to be worse than the disease.

NOTES

1. S. Denning, "How CEOs Became Takers, Not Makers," *Forbes.com*, August 18, 2014, http://www.forbes.com/sites/stevedenning/2014/08/18/hbr-how-ceos-became-takers-not-makers/; S. Denning, "From CEO Takers to CEO Makers: The Great Transformation," *Forbes.com*, August 20, 2014, http://www.forbes.com/sites/stevedenning/2014/08/20/from-ceo-takers-to-ceo-makers-the-great-transformation/.

2. M. C. Jensen, and W. H. Meckling, "Theory of the Firm: Managerial Behavior, Agency Costs, and Ownership Structure," *Journal of*

Financial Economics 3, no. 4 (October 1976): 305–360. They put forward the absurd view that the executive's motivation is solely pecuniary: In due course, this assumption became a horrifying self-fulfilling prophesy.

3. S. Denning has written extensively on this subject; see "Retirement Heist: How Firms Plunder Workers' Nest Eggs," *Forbes.com,* October 19, 2011, http://www.forbes.com/sites/stevedenning/2011/10/19/retirement-heist-how-firms-plunder-workers-nest-eggs/; "GE Discusses Retirement Heist," *Forbes.com*, October 21, 2011, http://www.forbes.com/sites/stevedenning/2011/10/21/ge-discusses-retirement-heist/Heist Part 3, *Forbes.com*, October 22, 2011, http://www.forbes.com/sites/stevedenning/2011/10/22/retirement-heist-part-3-ellen-schultz-replies-to-ge/; and "How Your Pension Got Turned into Scotch or Cheese," *Forbes.com,* April 22, 2013, http://www.forbes.com/sites/stevedenning/2013/04/22/sorry-about-your-pension-scotch-cheese-or-golf/.

4. See the discussion in Chapter 7 and in S. Denning, "The Surprising Reasons Why America Lost Its Ability to Compete," *Forbes.com*, March 10, 2013, http://www.forbes.com/sites/stevedenning/2013/03/10/the-surprising-reasons-why-america-lost-its-ability-to-compete/.

5. M. Friedman, *Capitalism and Freedom* (Chicago: University of Chicago Press, 1962), 14.

6. O. Lobel, "Companies Compete but Won't Let Their Workers Do the Same," *New York Times*, May 4, 2017, https://www.nytimes.com/2017/05/04/opinion/noncompete-agreements-workers.html.

7. L. Summers, "U.S. Economic Prospects: Secular Stagnation, Hysteresis, and the Zero Lower Bound," *Business Economics* 49, no. 2 (2014), http://larrysummers.com/wp-content/uploads/2014/06/NABE-speech-Lawrence-H.-Summers1.pdf. At first, Summers argued that the problem was one of demand. In a later article, he suggested that "supply-side chokes" were also part of the problem. See Tomas Hirst, "Larry Summers Admits He May Have Been Wrong on Secular Stagnation," *Business Insider*, September 9, 2014, http://www.businessinsider.com.au/larry-summers-admits-he-may-have-been-wrong-on-secular-stagnation-2014-9.

8. Denning, "The Surprising Reasons Why America Lost Its Ability to Compete." Models based on the "invisible hand of the marketplace"

and "perfect competition" are turning out to be obsolete mantras, rather than useful guides to action. There are currently 4 million unfilled jobs in the United States.

9. Friedman, *Capitalism and Freedom*, 14.

10. See Chapter 7 and Denning, "The Surprising Reasons."

11. Ibid.

12. Ibid.

13. S. Denning, "Milton Friedman and the Fallacy of Good Intentions," *Forbes.com*, August 1, 2013, https://www.forbes.com/sites/stevedenning/2013/08/01/milton-friedman-and-the-fallacy-of-good-intentions/.

9.

THE TRAP OF SHARE BUYBACKS

Share buybacks have become a kind of corporate cocaine that induces a temporary feeling of invincibility but masks weakness and vacuity.

In 1787, Empress Catherine II of Russia made an unprecedented six-month trip to Crimea, the "New Russia," with her court and some foreign ambassadors. The area had been devastated by war. Amid fears of new hostilities, the purpose of the trip was to impress on Russia's allies how prosperous the region had become after rebuilding the region and bringing in Russian settlers.

The Empress was accompanied on the trip by the official who had been responsible for rebuilding the region and bringing in Russian settlers. The official, Grigory Potemkin, happily combined this official role with his unofficial function of bedroom companion and lover of the sex-hungry Empress.

Potemkin's problem on the visit was that the reconstruction of the area was not as far along as desired. Not wanting to disappoint the wishes of his imperial mistress, and being an energetic fellow, he hit upon an ingenious scheme. Why go to the bother of generating actual prosperity when prosperity could be simulated?

Potemkin had a team of workers develop some portable villages. Prior to the arrival of the Empress's barge, Potemkin's men, dressed up as peasants, would show up at the site and assemble a village. At night, in the midst of the barren territory, the fake settlements with their glowing fires would comfort the Empress and her foreign entourage. Once the Empress's barge had departed, the village would be disassembled and rebuilt downstream for the imperial visit the next evening.

The stratagem was a personal success for Potemkin. The Empress was sufficiently pleased with his multiple services that he solidified his hold on power. For Russia, however, the outcome was less happy. The visiting ambassadors detected the difference between the real and the fake villages and Potemkin's deception was condemned by his political opponents. Shortly after the imperial visit, the region was plunged into a war between Russia and the Ottoman Empire.

Fast-forward a couple of hundred years and we can see how Potemkin's tactics have been adapted to the modern world of corporate finance through the use of share buybacks.

The man who brought the debacle to public notice is Bill Lazonick, a strong-minded economics professor who has been working for several decades on the role of innovative business enterprises in generating productivity. His work has focused on the growing financialization of the U.S. economy, taking into account broader historical and global perspectives such as the British Industrial Revolution, Japan, and China. He set out to construct a rigorous theory of economic growth, grounded in the microeconomics of the innovative enterprise. Lazonick has systematically challenged the faith of mainstream economists in the magic of the marketplace. Over time, he became a leading critic of mainstream economic thinking and current business practices.

In 1993, after the publication of three books in the three previous years, Lazonick made the unusual academic move of leaving a tenured position at an elite private university for one at a regional public institution, University of Massachusetts Lowell, so that he could have more freedom to pursue innovative thinking.

For several decades, Lazonick's work was known mainly among academic economists. But in 2014, a breakthrough came. Lazonick's work was featured in an article in *Harvard Business Review*. It presented the business world with astonishing news: Many major public corporations are engaged in buying back their own shares to an extent that constitutes stock-price manipulation on a macroeconomic scale.[2]

It was hard to argue with Lazonick's conclusions, which were based on decades of detailed data collection and economic analysis. Other mainstream journals picked up the theme. *The Economist* called share buybacks "an addiction to corporate cocaine." Reuters called it "self-cannibalization." The *Financial Times* called it "an overwhelming conflict of interest." In March 2015, Lazonick's article won the HBR McKinsey Award for the best *Harvard Business Review* article of the year.[3]

How had so many of the biggest and most respected companies in the world gotten involved in stock-price manipulation on such a massive scale? Why is it still tolerated by regulators?

It's simple, Lazonick explains. Once firms began in the 1980s to focus on maximizing shareholder value as reflected in the current share price, the actual capacity of these firms to generate real value for the organization and its shareholders began to decline as cost-cutting, dispirited staff, and limited capacity to innovate took their toll. Thus, the C-suite faced a dilemma. They had promised increasing shareholder value, and yet their actions were systematically destroying the capacity to create that value. What to do?

They hit upon a wondrous shortcut: Why bother to *create* new value for shareholders? Why not simply *extract* value that the organization had already accumulated and transfer it directly to the shareholders (including themselves) by way of buying back their own shares? By reducing the number of shares, firms could, through simple mathematics, boost their earnings per share. The result was usually a bump in the stock price—and short-term shareholder value.

Of course, by diverting important resources to boost the stock price, the tactic ran the risk of further hindering the firm's capacity to innovate and generate fresh value for customers in future. But why worry about that? With luck, by the time it became apparent that the firm had undermined its long-term capacity to add real value to customers, the executives responsible for the decisions would be safely retired, with bonuses already paid. The loss of capacity to create value would be someone else's problem.

There was just one snag. Jacking up the share price with large-scale share buybacks would constitute stock-price manipulation and hence would be illegal. But no problem! In 1982, the Reagan administration was happy to remove the impediment and the Securities and Exchange Commission (SEC) instituted Rule 10b-18 of the Securities Exchange Act.

Naturally, the SEC didn't announce that stock-price manipulation was being legalized. That would have created a political outcry. Instead, they passed a very complicated rule that made it seem that stock-price manipulation was still illegal, but provided protection to firms so that it would be hard to detect and almost invulnerable to legal challenge.

The complex rule that the SEC came up with is the following. Bear with me if you find the rule hard to understand. That of course is the whole point of the language: to make it hard to understand.

> Under the rule, a corporation's board of directors can authorize senior executives to repurchase up to a certain dollar amount of stock. After that, management can buy more company's shares provided that, among other things, the amount did not exceed a "safe harbor" of 25 percent of the previous four weeks' average daily trading volume. Since companies are not required to report daily repurchases, the SEC has no way of determining whether a company has breached the 25 percent limit without a special investigation. The rule preserves the illusion that stock price manipulation is still illegal. But in practice companies can repurchase their shares on the open market with virtually no regulatory limits. Even better: Since the share purchases are happening in the background, the public can't see what is going on.

And so the floodgates opened. The resulting scale of share buybacks is mind-boggling. Over the years 2006–2015, Lazonick's research shows that the 459 companies in the S&P 500 Index that were publicly listed over the ten-year period expended $3.9 trillion on share buybacks, representing 54 percent of net income, in addition to another 37 percent of net income on dividends. Much of the remaining 10 percent of profits are held abroad, sheltered from U.S. taxes. The total of share buybacks for all U.S., Canadian, and European firms for the decade 2004–2013 was $6.9 trillion. The total share buybacks for all public companies in just the United States for that decade was around $5 trillion.[4]

Theoretically, the SEC could launch investigations and intervene to prevent what is obviously share-price manipulation. But the SEC has been inactive. "The SEC," says Lazonick, "has only rarely launched proceedings against a company for using them to manipulate its stock price."

In fact, the SEC recently declared itself powerless to do anything about the problem. Thus, in July 2015, when Tammy Baldwin, the Democratic Senator from Wisconsin, asked the SEC head appointed by the Obama administration, Mary Jo White, to look into the issue of stock-price manipulation resulting from share buybacks, White replied that the SEC could not consider the issue because of the protection offered by Rule 10b-18. The prospects of changing that ruling in the current investor-friendly administration seem even more remote.[5]

True, not all share buybacks are bad. They can make sense "when the share price is—truly—below the intrinsic value of the productive capabilities of the company and the company is profitable enough to repurchase the shares without impeding its real investment plans."[6]

But these "constitute only a small portion of modern buybacks," says Lazonick. "Surgical interventions with private tenders are the exception." Most of the trillions of dollars in share buybacks have been made on the open market, often when the share price is high. Open-market share buybacks generally "come at the expense of investment in productive capabilities."[7]

Share buybacks not only continue on a massive scale. They are increasing. The *Financial Times* reported that in 2015, U.S. companies unleashed "a share buyback binge. . . . Now the market stands on the cusp of seeing a record of more than $1 trillion returned to shareholders in the form of dividends and stock repurchases this year." This is happening at a time when share prices are at record highs.[8] (See Figure 9-1.)

The practice of share buybacks has been further enabled by the actions of the central bank since 2008 in providing cheap money in almost unlimited quantities to large corporations via the banking system. Big corporations could thus borrow vast amounts of money at practically no cost and use the loans to fund stock buybacks for the benefit of the shareholders and their executives.

The ostensible goal of the central bank's action was to stimulate the economy out of the Great Recession. But now, nine years later, low-cost money is still available and the economy continues to sputter along in a state of secular economic stagnation.

Although the poet Wallace Stevens, in a moment of fancy, compared money to poetry, money is more often like information. When money can be obtained in large quantities at zero interest rates over long periods, it fosters the illusion that money is limitless, which leads both governments and businesses astray. The principal beneficiaries of the central bank action have been, not the population as a whole, but rather those who own shares. The stock market soars, and traders and the owners of assets exult. The result, however, is Potemkin prosperity, not the real thing.

Principal beneficiaries of share buybacks include activist shareholders. For instance, in recent years, activist hedge funds, who played absolutely no role in the success of firms over the decades, have purchased large amounts of their stock and then pressured the companies to implement huge buyback programs. The transactions transferred vast amounts of resources to the activists with no gain to the real economy.

Rather than investing resources in creating new value for customers with market-creating innovations, share buybacks are siphoning off resources to hand back to shareholders. The suggestion that the siphoned-off resources are being used elsewhere in the economy to create jobs stretches credibility. It ignores the reality that the resources are mostly deployed to finance other value-extraction schemes.

Share buybacks, Lazonick points out, ignore the legitimate claims of "other participants in the economy who bear risk by investing without a guaranteed return. . . . As risk bearers, taxpayers, whose dollars support business enterprises, and workers, whose efforts generate productivity improvements, have claims on profits that are at least as strong as the shareholders."[9]

In effect, many public companies have become giant "reverse Ponzi schemes." Thus, a Ponzi scheme attracts investments into a firm on the false premise that it is a valuable company. By contrast, a reverse Ponzi scheme takes a valuable firm and systematically extracts value from it. The firm appears to be making profits even as it systematically destroys its own earning capacity by handing over resources to shareholders.

The Challenge for Public Policymakers

The systematic extraction of value from corporations on a macroeconomic scale isn't an issue of a few misguided individual CEOs or occasional aberrations from the norm. It's one of fundamental institutional failure. CEOs are extracting value from their firms and helping other CEOs do the same thing. Boards are giving the C-suite incentives to do it. Business schools are teaching them how to do it. Institutional shareholders have been complicit in what the CEOs are doing. Regulators search for individual wrongdoers, usually those below the C-suite, while remaining blind to systemic failure. Central bankers indirectly fund the operation and close their eyes to the economic consequences. See Figure 9-2.

It is time for society's leaders to fix a flawed system. A systemic solution goes well beyond the obvious step of repealing the "safe harbor" provided by Rule 10b-18 of the Securities Exchange Act, which effectively enables big corporations to pursue stock price manipulation on a massive scale. Resolving the problem involves many institutions: Corporations, corporate boards, investors particularly institutional investors, legislators, regulators, business schools, and central banks all need to think and act differently.

The change must start with Wall Street itself. As of mid-2017, there were some signs that this was already happening. Investors now are punishing companies that have focused on share buybacks. "Infatuation with buybacks has ended for both companies and investors," David Kostin, Goldman's chief U.S. equity strategist, said in a note to clients. "Experience shows that firms repurchasing shares at extremely high valuations regret those actions when the stock price inevitably de-rates."[10] (We will come back to these issues in Chapter 12 and Box 12-2.)

The Challenge for Agile Leaders in Dealing with the Stock Market

The more pressing challenge for Agile leaders is how to deal with continuing pressures to shift resources from investment into share buybacks. The answer lies in recognizing that going along with these pressures is a choice, not a necessity. Thus, some firms have made it explicit from the outset that they will not be playing the value-extraction game. Amazon is the most prominent example. It never focused on short-term shareholder value. At Amazon, shareholder value is the result, not the operational goal. Amazon's operational goal is market leadership. Although its short-term profits have been variable, the stock market has handsomely rewarded Amazon's long-term strategy.

"We first measure ourselves," says Amazon's CEO, Jeff Bezos, "in terms of the metrics most indicative of our market leadership: customer and revenue growth, the degree to which our customers continue to purchase from us on a repeat basis, and the strength of our brand. We have invested and will continue to invest aggressively to expand and leverage our customer base, brand, and infrastructure as we move to establish an enduring franchise."[11]

Amazon obsesses over customers, not shareholders. "From the beginning, our focus has been on offering our customers compelling value." Long-term shareholder value, says Bezos, "will be a direct result of our ability to extend and solidify our current market leadership position. The stronger our market leadership, the more powerful our economic model. Market leadership can translate directly to higher revenue, higher profitability, greater capital velocity, and correspondingly stronger returns on invested capital."[12]

And it is not impossible to commit to customer value in midstream. For instance, on his first day as CEO of Unilever, Paul Polman warned his shareholders that he was not going to maximize shareholder value as reflected in the stock price. "Immediately, the Dutch-born Polman put his shareholders on notice," writes Forbes contributor Andy Boynton. "He declared that they should no longer expect to see quarterly annual reports from the company, along with earnings guidance for the stock market. Unilever, he explained, was now taking a longer view. CEO Polman went a step further, urging shareholders to put their money somewhere else if they don't 'buy into this long-term value-creation model, which is equitable, which is shared, which is sustainable. I figured I couldn't be fired on my first day,' Polman later said. Unilever's share price initially sank, but eight years later, Polman is still the CEO. Its stock price has soared and Unilever is prospering by delivering real value to real customers in a sustainable way."[13]

A key step is to educate investors that a focus on short-term returns and the wholesale use of share buybacks actually destroys shareholder value. Amazon, Unilever, and Berkshire Hathaway have made clear to investors that they won't be playing the shareholder value game. Instead of being punished for ignoring the short term, these firms have been rewarded with strong investor support.[14] See Figure 9-3.

Much of the pressure to manage for the short term is thus self-inflicted. To a large extent, business leaders get the investors that their behavior attracts. The first step in emancipating their shareholders from indulging in "corporate cocaine" is to emancipate themselves from it.[15]

The Challenge for Agile Managers Within the Corporation

The pressures within public corporations to devote resources to share buybacks can pose serious headwinds on Agile management. While one part of the corporation may be doing its utmost to add value for customers through Agile management, another part—the C-suite—may be extracting resources to buy back shares in response to pressures from the stock market.

When leaders succumb to these pressures, corporations become schizophrenic. As more resources are shifted into share buybacks, there are insufficient resources to support investment in innovation. And because there is insufficient innovation, there is a need for more share buybacks. In a world in which boosting the current stock price is the overriding concern, investments in market-creating innovations for customers are often what gets cut.

For the Agile manager in a public corporation, the challenge is to see what can be done to insulate the firm's Agile journey from these pressures. The first step is for managers to educate themselves on the nature and extent of the problem. Many Agile managers are oblivious to what is going on in the organization beyond their immediate sphere of influence. Then they are surprised when an edict comes down from the C-suite shifting resources away from needed innovation in an effort to boost the current stock price.

As a practical matter, Agile managers must face the reality that unless they can wean the C-suite from the "corporate cocaine" of share buybacks and disabuse them of the merits of shareholder value thinking, the life of an Agile transformation within the firm is unlikely to be happy or long. The stakes are high: Either Agile management will take over the whole organization, or shareholder value thinking will crush Agile management.

Agile managers need to make the case that if the firm takes care of customers, shareholders will also do much better. The opposite simply isn't true: If a firm focuses on taking care of shareholders in the short term, customers don't benefit and, ironically, the gains for shareholders are usually ephemeral. It's the real market of providing goods and services to real customers that creates a sustainable future in a way that exploiting short-term financial opportunities can never achieve. Agile management thus produces meaning and motivation for organizations along with a real future. Agile managers must educate the C-suite that "corporate cocaine" may feel good in the short term but it has horrendous consequences for the firm's health.

If Agile managers are to win these battles within the organization, they must also understand and deal with the cost-oriented economics through which shareholder value thinking has become embedded in day-to-day decision making. What's involved in doing that is the issue to which we now turn in Chapter 10.

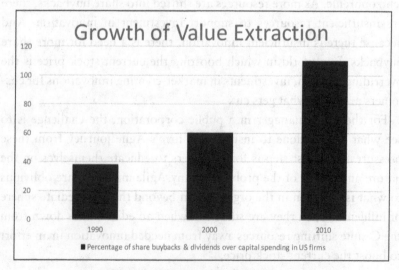

Figure 9-1. Growth of share buybacks.

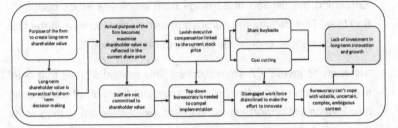

Figure 9-2. The vicious cycle of value extraction.

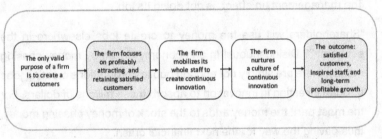

Figure 9-3. The virtuous circle of long-term value creation.

BOX 9-1

DEFENDING SHARE BUYBACKS

Executives give five main justifications for open-market share buybacks: "We're helping shareholders who own the company." Or, "We're preventing shareholder dilution." Or, "We're buying when prices are low to strengthen the company." Or, "We have run out of good investment opportunities." Or, "The shareholders will use the money to create jobs." None of these justifications are plausible.

♦ "Creating value for shareholders"? Wrong! Most open-market share buybacks provide temporary wins but systematically kill long-term value for shareholders. They are extracting value, not creating it.

♦ "Signaling confidence in the company's future"? How can this be so, when, as Lazonick points it, "over the past two decades major U.S. companies have tended to do buybacks in bull markets and cut back on them, often sharply, in bear markets"? What sort of a game is it when executives "buy high and, if they sell at all, sell low"? The answer is clear: share-price manipulation.

♦ "Offsetting the dilution of earnings from employee stock options"? Bad idea! This defeats the purpose of using stock options in the first place—namely, to encourage long-term performance.

♦ "Nowhere to invest"? A smokescreen! This pretext signals that chief executives are not performing their principal function of discovering new investment opportunities. In effect, executives are setting aside the hard work of creating sustained innovation and simply making money for themselves and their colleagues with the stroke of a pen. Top management, in effect, is not doing its job.

♦ "Shareholders will use the money to create jobs elsewhere in the economy"? Disingenuous! In an economy driven by money chasing money, there are simply not enough public corporations creating jobs through investing in innovation to absorb these trillions of dollars. For the most part, the money adds to the stock of money chasing money, thus paving the way for the next financial crash.

Practices like share buybacks generate inauthenticity in executives, filling their world with encouragements to suspend moral judgment. They receive incentive compensation to which the rational response is to game the system. And since they spend most of their time trading value rather than building it, they lose perspective on how to contribute to society through their work. Customers become marks to be exploited and employees become disposable resources.

Enlightened self-interest dictates that firms break out of the vicious cycle of value extraction (Figure 9-2) and embrace a virtuous circle of value creation (Figure 9-3).

10.

THE COST-ORIENTED ECONOMICS TRAP

Saving on expense may be an upside of tech innovation and invest-ment, but the real goal should be creating value.

—JOHN MCMANUS[1]

"The numbers give us no choice," the CFO declares. "It's obvious that if labor costs in another country are very much lower, it will be cheaper to outsource the work there, even if the workers there are somewhat less productive." Agile managers in public corporations are often confronted with such declarations. The argument can be even more compelling in software development than manufacturing because in a virtual world, the reasoning goes, geographical location doesn't matter: Software can be shipped in an instant from wherever it is developed to wherever it is needed. "The numbers are clear—we must offshore."

But where do the CFO's numbers come from? And what are their underlying assumptions? If Agile managers come to such discussions with no more than the self-evident premises of the Agile Manifesto, it will be like coming to a gunfight without a weapon. To prevail in such settings, Agile managers need a certain amount of understanding of financial accounting and its tools, the assumptions underlying the numbers, and their relationship to Agile management.

Agile managers need to understand the preoccupation of traditional economics with cutting costs. Ronald Coase won the Nobel Prize in Economics for his 1937 article explaining that firms exist because they reduce transaction costs.[2] The theory made sense when products were essentially commodities. But as the economy steadily shifts from commodities to complex differentiated or personalized products and services, and as power in the marketplace shifts from seller to buyer, adding more value to customers at lower cost has become steadily more important than internal efficiency.

The phenomenon is particularly striking in software. A few decades ago, when firms were dealing with big and slow-moving monolithic systems, it made sense to focus on efficiency. Today, as unit costs approach zero, storage space is almost infinite, and systems operate and interact almost instantaneously, competitive advantage rests almost entirely in what value can be added.[3]

Yet CFOs continue to emphasize cost-cutting, because it can appear to improve short-term profits and so boost the current share price. CFOs often come to the table with proposals for short-term gains that are not in the long-term interests of the corporation or its shareholders. Consider for instance the sad story of Dell Inc.

The Case of Dell Inc.

Dell is a multinational computer technology company based in Round Rock, Texas, that develops and sells computers. One important part of its story is told in *The Innovator's Prescription*, by Clayton Christensen and colleagues.[4] Several decades ago, Dell discovered that it could lower the cost of production by asking a Taiwanese electronics manufacturer, ASUSTeK, to make simple circuit boards inside a Dell computer. As a public corporation, Dell could go to Wall Street and show how offshoring would lower Dell's operating costs and remove some manufacturing from its balance sheet. Both Dell and ASUSTeK were happy with the arrangement. It was a win-win for both firms.

Then one day, ASUSTeK came to Dell with an interesting proposition. They said in effect: "We've been doing a good job making these little boards. Why don't you let us make the motherboard for you? Circuit manufacturing isn't your core competence anyway and we could do it for much less."[5]

Dell accepted the proposal because it made sense from a perspective of making money for shareholders in the short term: Dell's revenues were unaffected and its profits improved significantly, as it could save money by eliminating the staff who had previously worked on the task. On successive occasions, ASUSTeK came back and took over the motherboard, the assembly of the computer, the management of the supply chain, and the design of the computer. In each case, Dell accepted the proposal because Dell's revenues were unaffected and its profits improved significantly, as it cut back on the costs of its own staff. However, the final time ASUSTeK came back, it wasn't to talk to Dell. It was to talk to retailers and tell them that they could offer their own brand or any brand PC for 20 percent lower cost. As *The Innovator's Prescription* concludes:

> Bingo. One company gone, another has taken its place. How did it happen? There's no stupidity in the story. The managers in both companies did exactly what business school professors and the best management consultants would tell them to do—improve profitability by focusing on those activities that are profitable and by getting out of activities that are less profitable.[6]

This sequence of decisions, the authors say, made sense from Dell's point of view because the return on each of the decisions on outsourcing was high. At each step, Dell's revenues were unaffected and its profits improved significantly. The end result? ASUSTeK became Dell's formidable competitor, while Dell was reduced to hardly more than a brand, having lost the technical expertise it needed to innovate and grow its business.

The good news in the story is the sequel: In the end, Dell didn't quite die. But it had to take itself private, and fundamentally reinvent itself, in order to get back to delivering high-quality products that delight customers. The move is having some success. I can even report that when I had to replace my PC recently, I surveyed the market and discovered that the PC best corresponding to my particular needs was made by Dell. So Dell didn't die, but its near-death experience with offshoring was a costly one.

The widely accepted implication of the story is that the calculations made by Dell "dictated" that it offshore manufacturing. Yet a closer look at the numbers shows that this just isn't so. If Dell's managers had examined more closely what was involved in assuring Dell's future—by continuing to delight its customers—they would have realized that not only would they be steadily losing their capacity to grow the business, but they would actually be killing the business. Contrary to what Christensen and his colleagues say, there *was* stupidity in the story. Dell was *not* practicing "good management."

Given hindsight, we can see that Dell should have included in its calculations the costs and risks of running long-distance supply chains in foreign countries; the likely closing of wage differentials between the countries; the declining proportion of labor in the total costs of production given increased automation; the cost of knowledge and expertise being lost; and the cost of creating a competitor (ASUSTeK) that could make a better product at lower cost and so lower the margins that Dell would be able to charge in the future. The cost of crippling its future business was enormous, and it was missing from the calculation.

The calculation of costs, risks, and benefits based on a narrow view of costs, particularly labor costs, assumes that the firm's ongoing business will continue and grow and that any other costs are insignificant. If Dell had correctly assessed the full opportunity costs and risks of its actions, the calculation would have revealed the business disaster that was unfolding.[7] While analysts sometimes blame the tool being used for the outcome, they should rather blame the mindset applying the tool, which fell into the trap of cost-oriented economics.

Mastering the arduous task of calculating rates of return might seem tedious to Agile managers when compared to the thrill of building cool new products that delight customers. But often, it's in these tedious details that the battle for the future of Agile management is being lost—or won. If Agile managers don't make the effort to understand the nitty-gritty of the numbers and the financial analysis, they will be easily bulldozed by those who claim: "The numbers make us do it!" and, "Numbers don't lie!"

Let's also keep in mind that the problem with numbers and numbers-people isn't numbers. Traditional management was wrong about many things, but it was right about one thing: the importance of measurement. Peter Drucker repeatedly said that it was hard to improve something if you can't measure it, while also noting that in some areas, reliable measures were lacking.[8] The main trouble with measurement comes when managers measure the wrong thing. The problem with bean counting isn't the act of counting, but rather the mistake of counting beans. Managers need to be counting the elements that drive the business—customer outcomes, not merely outputs—and also using their judgment where good measures are lacking.

Here, as in other areas, what's key is the management mindset. A mindset that is set on extracting value from a corporation for the shareholders will come to different conclusions from the numbers, compared to a mindset focused on creating monetizable value to customers.

The Urge to Offshore

Take the decisions of U.S. public corporations on issues of offshoring over the last several decades. Whole industries have been shipped overseas in pursuit of short-term cost savings. For the last several decades, it became a lemming-like rush.

As one Fortune 500 executive told me:

> The decision to go to China was driven almost exclusively by the financial analysts, who told management they couldn't support a "buy" recommendation if the firm had no plans to be in China. Since the executives who make the decision are compensated with hefty stock options, it wasn't long before corporate was offering Mandarin lessons. These people understand the long-term implications of these decisions perfectly well, and they're very sympathetic to the poor b*****ds who will come after them and have to sort this mess out—after their options have been exercised, of course. The unholy alliance between the analyst community and executives heavily incentivized by stock options—given cover by the simplistic *apologia* of cost accountants who coo about labor savings in China but forget about the six-figure travel expenses—helps explain why manufacturing has left.

Incorporating the full cost of offshoring is still a pervasive issue in business today. In analyzing proposals, firms must get beyond rudimentary cost calculations focused on the cost of labor and instead consider the total cost and risk of extended international supply chains over time. There's an app for that: It's called the Reshoring Initiative. It has a website with an analytical tool enabling companies to calculate the full risks and costs of offshoring. It's called the Total Cost of Ownership Estimator.[9] And happily, it's free.

The Estimator poses a series of questions. What's the price of the part from each of the destinations? How far away is it? How often are you going to travel to see the supplier? How much intellectual property risk is there? How long do you think you are going to make it? It uses the answers to calculate twenty-five different costs that add up to the Total Cost of Ownership.

"Many companies that offshored manufacturing didn't do the math," Harry Moser, an MIT-trained engineer and founder of the Reshoring Initiative, told me. "A study by the consulting company, Archstone, showed that 60 percent of offshoring decisions used only rudimentary cost calculations, maybe just price or labor costs, rather than something holistic like total cost. Most of the true risks and costs of offshoring were hidden."[10]

"Often what firms find," says Moser, "is that whereas the offshoring price is perhaps 30 percent less than the U.S. price, all these other costs add up to more than 30 percent. If they are willing to recognize all of them, then they can see that it may be profitable to bring the work back."[11]

Why did so many smart managers make the same mistake? It's not a lack of mathematical capability. Their decision making was driven by the rush to generate short-term profits—or at least *appear* to be generating them—so as to raise the stock price.

A contributing factor was the belief that offshoring manufacturing to emerging markets would eliminate labor problems that were arising because of the top-down bureaucracy in U.S. firms. By offshoring production, firms were not only seeking to lower their production costs, they were also checkmating unions by removing jobs from this country. These considerations have led many managers to ignore the cliff they are driving their firms over.

A Permanent Loss of Expertise

Sadly, the cost of offshoring goes beyond the consequences for individual companies. Decades of such decision making have led to massive outsourcing of manufacturing and left U.S. industry without the capability to invent or manufacture the next generation of high-tech products that are key to growing the economy, as noted by Gary Pisano and Willy Shih in their classic article "Restoring American Competitiveness."[12] The authors showed how the United States lost its ability to develop and manufacture a slew of high-tech products, in many cases using technology that had been invented in America. For instance, Amazon can't manufacture a Kindle in the USA today, even if it wanted to. The expertise no longer exists in the United States.[13]

Pisano and Shih continue:

> So the decline of manufacturing in a region sets off a chain reaction. Once manufacturing is outsourced, process-engineering expertise can't be maintained, since it depends on daily interactions with manufacturing. Without process-engineering capabilities, companies find it increasingly difficult to conduct advanced research on next-generation process technologies. Without the ability to develop such new processes, they find they can no longer develop new products. In the long term, then, an economy that lacks an infrastructure for advanced process engineering and manufacturing will lose its ability to innovate.

Pisano and Shih gave a frighteningly long list of technologies "already lost"—some of them invented in the United States—along with a list of industries "at risk" that is even longer and more worrisome.[14]

Yet some firms are coming back to the States. For instance, Charles Fishman reports in *The Atlantic* that GE spent some $800 million to reestablish manufacturing in its giant facility—that was almost defunct—at Appliance Park in Louisville, Kentucky.[15] In 2012, GE opened an all-new assembly line to make cutting-edge, low-energy water heaters and high-tech French-door refrigerators.[16]

And a funny thing happened when GE decided to bring manufacturing of its innovative water heater back from the "cheap" Chinese factory to the "expensive" Kentucky factory. "The material cost went down. The labor required to make it went down. The quality went up. Even the energy efficiency went up. GE wasn't just able to hold the retail sticker to the 'China price.' It beat that price by nearly 20 percent."[17]

Fishman also reported that "time-to-market has also improved, greatly. It used to take five weeks to get the GeoSpring water heaters from the factory to U.S. retailers—four weeks on the boat from China and one week dockside to clear customs," he writes. "Today, the water heaters—and the dishwashers and refrigerators—move straight from the manufacturing buildings to Appliance Park's warehouse out back, from which they can be delivered to Lowe's and Home Depot. Total time from factory to warehouse: 30 minutes."[18]

Bringing industries back home is also leading to important discoveries. Thus, GE's water heater was designed in Louisville, but made in China. When GE took over the manufacturing, it discovered that its own designs involved a tangle of copper tubing that was difficult to weld together. "In terms of manufacturability, it was terrible," Fishman noted one GE executive saying.

So, GE's designers got together with the welders and redesigned the heater so that it was easier and cheaper to make. By having those workers right at the table with the designers, the work hours necessary to assemble the water heater went from ten hours in China to two hours in Louisville.

"For years," Fishman writes, "too many American companies treated the actual manufacturing of their products as incidental—a generic, interchangeable, relatively low-value part of their business. If you spec'd the item closely enough—if you created a good design, and your drawings had precision; if you hired a cheap factory and inspected for quality—who cared what language the factory workers spoke? . . . It was like writing a cookbook without ever cooking . . . there is an inherent understanding that moves out when you move the manufacturing out. And you never get it back."[19]

What is only now dawning on American companies, Lou Lenzi, head of design for GE appliances, told Fishman, is that when you outsource the making of the products, "your whole business goes with the outsourcing."[20]

Explaining Agile Management to a CFO

While CFOs have certain weaknesses, lack of confidence in cost-oriented economics typically isn't among them. Prudent Agile managers need to come to discussions with CFOs and their staff with as much advance information as possible about the CFO's mindset and should exercise caution in attacking fundamental assumptions up front.

Agile managers should thus begin from a thorough understanding of the responsibilities of a CFO—and their legal limits. CFOs are required by law to carry out financial accounting in accordance with generally accepted accounting principles (GAAP) issued by the Financial Accounting Standards Board. These rules are enforced by the Securities and Exchange Commission (SEC) and other local and international regulatory agencies and bodies. CFOs have no option but to implement those rules. However, the laws also have limits: For instance, there is no legal requirement to apply the rules of cost accounting.

Within the overall financial accounting framework, CFOs have some latitude in the choice of the accounting system that forms the basis for the required periodic financial reports. The accounting systems of the different companies, and sometimes even different parts of the same company or organization, can thus vary somewhat.

Many firms embrace cost accounting and its preoccupation with identifying and reducing costs, particularly labor costs. Individual CEOs and managers may do their best to support Agile management, to encourage innovation, and to go the extra mile for the customer. But the powerful undertow of the cost-accounting framework creates a continuing risk that CFOs will be missing the point of Agile management, whose principal goal is adding value at lower cost, not just cutting costs.

Throughput Accounting

The challenge here is to get a traditionally minded accountant to start thinking beyond cost accounting and focus on the corporation's true purpose of generating value to customers and end-users at lower cost. In many situations, Agile management results in more value from less work and hence ultimately lower costs. However, if the CFO is not paying attention to value, the savings may be hidden.

Many accountants agree that cost accounting itself is problematic. But in a slow-moving profession like accounting, it has proved difficult to reach agreement on an alternative accounting system.

A step toward such a system was developed by Eliyahu Goldratt for managing factories and is known as *throughput accounting.*[21] It's a way for an organization to keep track of the velocity at which products and services move through an organization.

Throughput accounting isn't a total solution for Agile management, but it's an improvement on cost accounting. It can be used by a firm in its internal accounting to incorporate customer value, even though for external reporting, a public corporation will still have to follow generally accepted accounting principles.

If the CFO understands throughput accounting, an Agile manager is on the way to having a sensible discussion about the value of lean and Agile management. The CFO will already understand the importance of thinking through what adds value to the customer, not merely, "How can I cut costs?"

The follow-on conversation can be about the cost of work in process, the risk of delay, the importance of establishing a steady flow, and the customers' unmet needs. You might draw a chart that compares iterative work cycles to work carried in a traditional "waterfall" fashion, as shown in Figure 10-1.

Agile management provides value to clients at the end of each short cycle. By contrast, the waterfall approach proceeds according to a single elaborate plan, including all the specifications. Steadily mounting costs are incurred in the expectation that when all the pieces come together, the investment will pay off.

Even a traditionally minded cost accountant should be able to see from Figure 10-1 the risk involved in spending more and more with no interim return. And unless the environment is totally static, and the work totally predictable, the risks of a delay or a technical glitch occurring are significant. With so much invested with no return, any delay can be disastrous. Earlier delivery of value improves revenue and reduces risk, as shown in the example from Ericsson discussed in Chapter 1.

Moreover, Agile management focuses effort on the features, products, and services that customers and end-users actually use and value,

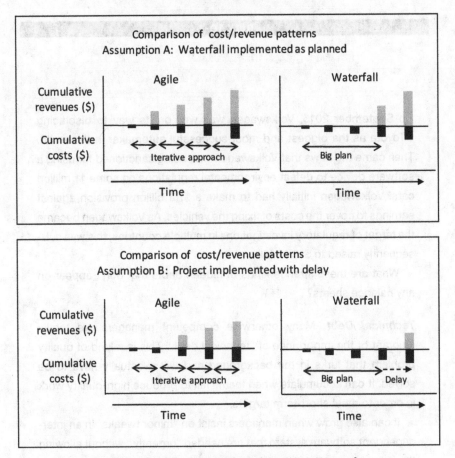

Figure 10-1. Comparison of cost/revenue patterns of Agile and waterfall implementation.

as opposed to features that are included in specifications "just in case" and that users may never in practice get around to using. In this way, Agile systematically reduces costs by eliminating low-value work.

Thus, even accepting the perspective of cost-oriented economics, a powerful case can be made for Agile management over the traditional waterfall approach. Nevertheless, victories of this kind will remain isolated and tactical unless and until there is a deeper understanding of the shift from a cost-orientation to Agile's value-driven perspective.

BOX 10-1:

TECHNICAL DEBT, REGULATORY DEBT, AND BRAND DEBT

In September 2015, Volkswagen was well on its way to displacing Toyota as the biggest and most successful automaker in the world. Then came the news that Volkswagen was being sanctioned for using a software device to defeat environmental regulations on some 11 million cars. Volkswagen initially had to make a $7.3 billion provision against earnings to cover the costs of fixing the vehicles. As Volkswagen became the target of regulatory investigations in multiple countries, this was subsequently raised to $18.32 billion.[1]

What are these strange kinds of hidden debts that don't appear on any balance sheets?

Technical Debt. Many otherwise competent managers are totally ignorant of the importance of "technical debt." This is a kind of quality problem that lurks in the background and will eventually have to be solved. It can accumulate when teams fail to produce high-quality work in order to meet short-term targets.

It can also grow when managers insist on "minor tweaks" in an interdependent software system that are needed "urgently," without allowing the developers to take the time to reconcile the changes with the system's architecture. Top management often becomes aware of the issue only when there is an abrupt system outage. That happens because when one piece of the system goes down, it can cause breakdowns in other subsystems, because of tight dependencies. The eventual result can be a total system outage, and the whole business can suddenly come to an abrupt halt.

These disasters can be important opportunities to educate top management on the implications of making changes to an interdependent system without in effect paying to do them properly. Cutting corners to achieve a performance goal is usually a fool's game. The firm can pay as it goes, or it can pay much more, later.

When these disasters emerge, they can be an opportunity for top management to reconsider its short-term perspective and learn the long-term costs of not doing things right in the first place. Management can learn how the accumulation of technical debt leads to increased support costs, difficulty in adding new features, unwanted complexity, and ultimately complete system breakdown.

Regulatory Debt. Taking shortcuts with software can also have massive regulatory implications. In some countries, system failure can involve penalties, even criminal liability and jail time for those responsible. In the case of Volkswagen, the regulatory penalties for deliberately flouting environmental regulations will be many billions of dollars. But that's not all.

Brand Debt. The cost of fixing the system and paying any fines may be only small pieces of the eventual damage suffered by Volkswagen. A study by Professors Jonathan Karpoff, D. Scott Lee, and Gerald S. Martin in the *Journal of Financial and Quantitative Analysis* shows that the cost in terms of loss of customer confidence can be many times higher.[2] The study of 585 firms, which had all been subject to enforcement action, showed that the expected loss in the present value of future cash flows due to lower sales and higher contracting and financing costs was over 7.5 times the sum of all penalties imposed through the legal and regulatory system.

A consideration of the Volkswagen case shows why. Volkswagen admits that a software switch was used to detect when the car was being tested. When it detected a test, it activated environmental controls. Under normal driving conditions, the switch activated a different engine calibration, which resulted in ten to forty times the pollution allowed under U.S. law.

Volkswagen wasn't doing this because it hated the environment. The engine calibration for normal driving dramatically increased performance and fuel efficiency. In effect, Volkswagen's ads, which trumpeted the combination of environmental friendly and powerful engines, were deceptive. The engines could do one or the other but not both. If the cars had been sold to operate in normal conditions with the environmentally sound calibration, the cars would have had much less driving oomph and they would have been much more costly to run.

Firms and investors thus need to be aware that hidden technical, regulatory, and brand debts may be more important than the financial debts disclosed on the firm's balance sheets.

NOTES

1. S. Denning, "Volkswagen and Its Hidden Debts, *Forbes.com*, September 24, 2015, http://www.forbes.com/sites/stevedenning/2015/09/24/volkswagen-and-its-hidden-debts/.

2. "Can Volkswagen Pass Its Emission Test?," *MarketWatch*, September 22, 2015, http://www.marketwatch.com/story/can-volkswagen-pass-its-emissions-test-2015-09-22; J. M. Karpoff, D. S. Lee, and G. S. Martin, "The Cost to Firms of Cooking the Books," *Journal of Financial and Quantitative Analysis* 43 (September 2008): 581–612, http://papers.ssrn.com/sol3/papers.cfm?abstract_id=652121.

BOX 10-2

U.S. VS. GERMAN MANUFACTURING

Although the loss of jobs through offshoring is often presented as the inevitable consequence of global changes in relative labor costs, studies show that different countries have been affected to different degrees. Thus between 2000 and 2009, the United States lost 33 percent of its manufacturing jobs, while Germany only lost 11 percent.[1] Why?

Like the United States, Germany has generally played by the rules of the global trade system. But unlike the U.S., many German firms are privately owned. They rarely emulated U.S. firms in the pursuit of maximizing shareholder value as reflected in the current stock price. Private owners of firms could see that a sharp focus on short-term profits hindered the creation of true long-term wealth. They had no interest in financial engineering gadgets, like share buybacks, to make the firm's shares appear to be doing better than they really were.

Germany also made efforts to create a context for innovation throughout the 2000s, steadily bolstering the competitiveness of its manufacturing sector. "Germany set about enacting a range of comprehensive economic reforms to increase the competitiveness of Germany's economy throughout the 2000s, including making its tax code more competitive, increasing investment in apprenticeship programs, increasing investment in industrially relevant applied R&D, and during the Great Recession introducing the short-time work program rather than firing workers outright as often happened in the United States. . . . And Germany is not alone; many more of America's competitors—including Japan, Korea, Holland, Taiwan, and even China—worked feverishly throughout the 2000s to bolster their science, technology, and innovation ecosystems that underpin the competitiveness and innovation potential of their private sector enterprises."[2]

The picture in the United States is very different. With their sharp focus on maximizing shareholder value, firms invested less in shared resources such as pools of skilled labor, supplier networks, an educated populace, and the physical and technical infrastructure on which U.S. competitiveness ultimately depends.[3]

In effect, while Germany and Asian competitors were strengthening their capacity to compete, U.S. firms were doing the opposite, and allowing the policy environment supporting the competitiveness of U.S. manufacturing industries to decline.

NOTES

1. A. Nager, "America's Job Loss Outpaces Other Leading Industrialized Countries," *Innovation Files*, August 19, 2014, http://www .innovationfiles.org/how-americas-manufacturing-job-loss-outpaces-other-leading-industrialized-countries/.
2. Ibid.
3. P. Porter, J. Rivkin, and R. M. Kanter, "Competitiveness at the Crossroads: Findings of Harvard Business School's 2012 Survey on U.S. Competitiveness," February 2013, www.hbs.edu/ competitiveness/pdf/competitiveness-at-a-crossroads.pdf.

11.

THE TRAP OF
BACKWARD-LOOKING STRATEGY

The future cannot be logically deduced from its past.

—JOHN DEWEY[1]

S trategy has been a major activity in business for decades. Why then didn't strategy reveal the disaster that was unfolding from maximizing shareholder value, or from the resort to the "corporate cocaine" of share buybacks, or from mass offshoring? Was it because strategy *couldn't* help? Or is the *practice* of business strategy inherently flawed? To help answer these questions, let's look at the sad story of the Monitor Group.

The Monitor Group was a strategy consulting firm cofounded in 1983 by the legendary business thinker Michael Porter. In November 2012, Monitor was unable to pay its bills and filed for bankruptcy protection. Why didn't the highly paid Monitor consultants use their own strategy analysis to save themselves?

After all, Monitor's demise hadn't happened like a bolt from the blue. The death spiral had been going on for some time. In 2008, Monitor's consulting work slowed dramatically during the financial crisis. In 2009, the firm's partners had to advance $4.5 million to the company and defer $20 million in bonuses. Then Monitor borrowed a further $51 million from the private equity firm Caltius Capital Management.

Beginning in September 2012, the company was unable to pay the monthly rent on its Cambridge, Massachusetts, headquarters. In November 2012, Monitor missed an interest payment to Caltius, putting the notes in default and driving the firm into bankruptcy.[2]

Was it negligence, like the cobbler who forgot to repair his own children's shoes? Had Monitor tried to implement its own strategy framework and executed it poorly? Or had Monitor implemented the strategy well but the strategy didn't work?[3]

■

The story of Monitor is a strange tale. It began in 1969, when Michael Porter graduated from Harvard Business School and crossed the Charles River to get a doctorate at Harvard's Department of Economics. There he learned that excess profits were real and persistent in some companies and industries because of structural barriers to competition. To the public-minded economists in the Department of Economics, the excess profits of these low-competition situations were an important problem to be solved.

Porter saw that what was a problem for the economists was, from a business perspective, a solution to be pursued. It was even a silver bullet. An El Dorado of unending above-average profits? Protected by permanent structural barriers? Exactly what business executives were looking for—a shortcut to fat city!

Why go through the risk and hassle of creating bold new products and services when the firm could simply position its business so that structural barriers ensured endless above-average profits?

Why not call this trick "the discipline of strategy"? Why not announce that a company occupying a position within a sector that is well protected by structural barriers would have a "sustainable competitive advantage"?

Why not proclaim that finding these El Dorados of unending profits would follow, as day follows night, by having highly paid strategy-analysts doing massive amounts of rigorous data collection and analysis? Which CEO would *not* want to know how to reliably generate perpetual profits? And why not set up a consulting firm that could satisfy that want?

■

And so it was that in March 1979, Michael Porter published his findings in *Harvard Business Review* in an article entitled "How Competitive Forces Shape Strategy." He followed it up the next year with his book *Competitive Strategy*.[4] These writings started a revolution in business strategy. Michael Porter became to the new discipline of business strategy "what Aristotle was to metaphysics."[5]

Better yet, the newborn profession of business strategy presented itself as a master discipline—the discipline that synthesizes all of the other functional subdisciplines of management into a meaningful whole. It was even presented as defining "the purpose of management and of management education."[6]

In 1983, Porter cofounded his consulting company, the Monitor Group, which over the succeeding decades generated hundreds of millions of dollars in fees from clients and provided rich livelihoods for other consulting firms, like McKinsey, Bain, and BCG.

Porter became "a giant in the field of competition and strategy," writes Joan Magretta in her 2012 book, *Understanding Michael Porter*: "Among academics, he is the most cited scholar in economics and business," she writes. "At the same time, his ideas are the most widely used in practice by business and government leaders around the world. His frameworks have become the foundation of the strategy field."[7]

There was just one hitch. What was the intellectual basis of this now vast enterprise of locating a firm's "sustainable competitive advantage"? Porter might have offered ways to create exciting new lines of business that would be difficult to compete against. Or he might have proposed management actions that could enhance profits from existing businesses. But unending above-average profits that could be deduced from the existing structure of the industry? Here we are in the realm of unicorns and phlogiston. Ironically, like the search for the Holy Grail, it was the very fact that the goal was mysteriously elusive that drove executives onward to continue the quest.

These strategic planning efforts entailed massive data gathering and analyses, as every conceivable facet of the competitive landscape was explored. But because there was no data on the future, the data gathering was inevitably about the past, from which extrapolations of the future were made. The data was also biased in another way, because there was no data on the unknown, only the known. This meant that possibilities beyond what was currently known about the industry didn't show up in the analyses.

The comfort that was gained from detailed analysis was thus inherently backward-looking. The possibility that the future might be very different from the past was masked by the seeming solidity and comprehensiveness of the data and the analyses that had been assembled. Thus, fast-growing and small competitors often didn't look like significant risks even when they began to make inroads. Unexpected moves by competitors were often absent from the analyses. Shifts in technology and customer attitudes and lifestyles were missed. The life expectancy of current businesses tended to be overestimated. Overall, the assurance that the future was secure—based largely on the thoroughness of the analyses of data from the past—was illusory.

Yet massive backward-looking strategy exercises continued. Talk of "rigorous analysis," "tough-minded decisions," and "hard choices" combined to hide the fact that there was no evidence that sustainable competitive advantage could be created in advance by studying the past structure of an industry. It didn't hurt the business of strategy consulting that the approach to strategy was—and still is—taught as a compulsory subject in business schools, leading to generations of executives who have been indoctrinated in this thinking.

Although the conceptual framework could sometimes shed light on excess profits in retrospect, it was generally unhelpful in predicting them in prospect. Backward-looking strategies are of course 100 percent accurate in hindsight, but in foresight, they miss the unexpected and the unforeseen. "The point is not that the strategists lack clairvoyance; it's that their theories aren't really theories—they are 'just-so' stories whose only real contribution is to make sense of the past, not to predict the future."[8]

■

Why had strategy gone astray? How had business managed to turn strategy into something that was no help in dealing with the future? Had Porter misconceived the very idea of strategy? Does Agile management have any need for strategy? To answer these questions, we need to look at the broader history of strategy.

Strategy is in fact a very old idea. It got an early start in the sixth century BC, with the appearance of the Chinese military treatise *The Art of War*, attributed to Sun Tzu, a high-ranking military general and strategist of the time. It consists of insights and maxims that still guide military leaders today:

- "War is a necessary evil that must be avoided whenever possible."
- "War should be fought swiftly to avoid economic losses."
- "Avoid massacres and atrocities because this can provoke resistance."

Nevertheless, a *systematic* consideration of military strategy only began in the late-eighteenth century as part of the Enlightenment belief in applying reason to all aspects of human activity. Strategy came to be seen as developing the perfect comprehensive plan on a map for aligning and deploying the armed forces. The actual execution of military operations was a matter of tactics for lower-level folk, who would, in theory at least, implement the plan as articulated at the top.

In practice, however, the gap between strategy and implementation created risks of miscommunication, and of a lack of commitment to implement the plan as conceived. Moreover, high-level strategy was likely to be devised without the vital insights of those with more specific and up-to-date knowledge of what was happening on the ground.

Over time, frustration with strategy so conceived led to the emergence in the nineteenth century of the German theorist Carl von Clausewitz, who argued in his famous (but unfinished) book, *On War,* that strategy is shaped by "a remarkable trinity—composed of primordial violence, hatred, and enmity."[9] Human action took place amid "the fog of war" and was affected by the "friction" from the unexpected interaction of different factors, in which rational planning could only play a limited role.

The implications of "the fog of war" and the "friction" that arose between conception and implementation were further developed by Helmuth von Moltke, who was appointed Chief of the Prussian (later German) General Staff in 1857. The dictum that made him famous was: "No plan of operations extends with any degree of certainty beyond the first encounter with the main enemy force."[10]

In some ways, von Moltke can be seen as the godfather of Agile management. To cope with uncertainty, von Moltke developed and applied the concept of *Auftragstaktik* (literally, "mission tactics"), a strategic approach stressing decentralized initiative within an overall strategic design. Von Moltke had no time for perfect comprehensive plans. He believed that, beyond calculating the initial mobilization and concentration of forces, leaders at all levels of the force needed to make decisions based on an assessment of a fluid, constantly evolving situation within an overall strategic design.

In the twentieth century, von Moltke's thinking grew steadily more influential in the military and in due course became the formal doctrine of the U.S. Army—at least on the battlefield. The Army in its formal theory of warfare thus contrasts information-based "detailed command" with action-oriented "mission command."[11]

Detailed command assumes that the world is deterministic, predictable, orderly, and certain, while *mission command* accepts that the world is probabilistic, unpredictable, disorderly, and uncertain.

Detailed command leads to centralization, coercion, formality, tight rein, imposed discipline, obedience, compliance, optimal decisions that take place later, and a focus on harnessing ability at the top—in a word, bureaucracy. By contrast, mission command is characterized by decentralization, spontaneity, informality, loose rein, self-discipline, initiative, cooperation, acceptable decisions that are made faster, and a focus on harnessing ability at all levels—in essence, Agile management.

The mission-command approach to strategy leads to a more flexible approach to operations, with a greater understanding throughout the organization and, overall, a more agile and effective organization.[12] In the twentieth century, the military steadily shifted from information-based strategy to mission-command strategy, particularly as warfare became more asymmetric.

■

Yet at the very moment that the military was abandoning top-down, information-based strategy for the conduct of its operations, strategists in business were embracing it. Strategy in business in the 1980s thus emerged within the conceptual framework of top-down, information-based strategy, with elaborate exercises to be primarily conducted at the top of the organization with unidirectional communications.

This approach might have been appropriate in a marketplace that was oligopolistic, stable, and predictable. But in business, precisely the opposite was taking place. Globalization, deregulation, and new technology were blowing away most of the barriers that had created the "enduring excess profits" that had so worried Harvard's Economics Department, and so thrilled the young Michael Porter, back in the 1970s.

Within the broader historical evolution of strategy, the approach was an aberration. The widespread adoption of information-based strategy in business, built on data from the past, helps explain the frequent outcome of strategic planning exercises to do "more of the same" (see Box 11-1). It also helps explain why firms consistently missed disruptive innovation.[13] It could hardly be otherwise when firms pursued information-based strategy built on data from the past.

But an information-based approach to strategy was not the only problem embedded in the approach to business strategy. Another was the very conception of strategy.

Porter began his publishing career with his 1979 article for *Harvard Business Review*, "How Competitive Forces Shape Strategy." The article was republished by *HBR* in 2008 with the proud and accurate boast that the article "has shaped a generation of academic research and business practice."[14]

The 1979 article starts with a very strange sentence: "The essence of strategy is coping with competition."[15] In effect, at the heart of Porter's concept of strategy is the idea that strategy is about protecting businesses from business rivals. The goal of strategy, business, and business education is to find a safe haven for businesses from the destructive forces of competition.

To accomplish this, Porter argued, a strategist must consider five forces: (1) the bargaining power of suppliers, (2) the bargaining power of customers, (3) the competitive rivalry among existing firms, (4) the threat of new entrants, and (5) the threat of substitute products (see Figure 11-1). The stronger any or all of these forces are, the more competitive the industry will be, and thus the lower the prospects for excess profits. The goal of strategy is to position the firm in a location where those forces are least operative. Therein lies the key to "sustainable competitive advantage."

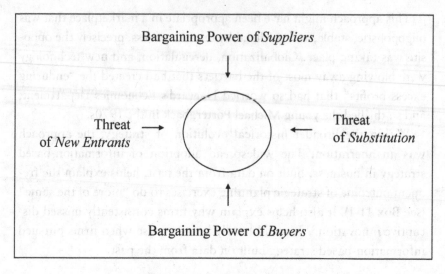

Figure 11-1. Porter's five forces.

Porter's article thus ignored Peter Drucker's foundational insight of 1954 that the *only* valid definition of business purpose is to *create a customer*. In Porter's competition-based model, customers are only relevant to strategy if they have bargaining power. The thinking underlying the five-forces notion of business strategy illustrates the pre-Copernican theory of management discussed in Chapter 3, in which the firm is the center of the commercial universe and the customer is a relatively unimportant thing on the periphery, if visible at all (see Figure 3-2).

Defining the essence of strategy as coping with competition also led to the misconception of business as a zero-sum game. As Porter wrote in his 1979 *HBR* article, "The state of competition in an industry depends on five basic forces. . . . The collective strength of these forces determines the ultimate profit potential of an industry."[16] On this basis, the ultimate profit potential of an industry is a finite fixed amount: The only question is which firm is going to get which share of it.

Sound business is, however, "unlike warfare or sports in that one company's success does not require its rivals to fail. Unlike competition in sports, every company can choose to invent its own game," says Joan Magretta. "A better analogy than war or sports is the performing arts. There can be many good singers or actors—each outstanding and successful in a distinctive way. Each finds and creates an audience. The more good performers there are, the more audiences grow and the arts flourish."[17]

What's wrong here is the concept of "the essence of strategy." The essence of strategy is not coping with competition—a contest in which a winner is selected from among rivals. The essence of business strategy is to add value for customers. Porter's five-forces model of strategy misses the basic point that ultimately customers are the determinant of business success.

The error in thinking that the purpose of strategy is to defeat business rivals rather than add value to customers has of course been aggravated by the epic shift in the power of the marketplace from the seller to the buyer, now that customers have options and reliable information about those options and an ability to communicate with other customers.

In the studies of the oligopolistic firms of the 1950s on which Porter founded his theory, it appeared that structural barriers to competition were widespread, impermeable, and essentially permanent. Over the following half century, the winds of globalization and the Internet blew away most of these barriers, leaving the customers in charge of the marketplace. Except for a few areas, like health and defense, where government regulation offers some protection, there are no longer any safe havens for business. National barriers collapsed. Knowledge became a commodity. New technology fueled spectacular innovation and rapid change. Entry into existing markets was startlingly easy. New products and new entrants abruptly redefined whole industries, killing old ones and creating new ones.

The "profit potential of an industry" turned out to be not a fixed quantity, with the only question being who would get which share, but rather a highly elastic concept, expanding dramatically at one moment or collapsing suddenly at another, with both competitors and innovations seemingly coming out of nowhere. Disruptive innovations destroyed company after company that believed in its own "sustainable competitive advantage." Just look at the diminished fortunes of once-flourishing category leaders such as Nokia, Kodak, Sony, Research in Motion, Motorola, HP, Borders, Circuit City, Sears, and JCPenney.

Or look at the massive businesses generated by Amazon, Apple, Facebook, Google, or Netflix when they provided new value to customers. These firms have not merely been extracting a larger share of "the profit potential of their industries" or "taking advantage of structural barriers in their industry." They have been *creating* vast new markets.

There is thus a straight line from the conceptual error at the outset of Porter's writing to the debacle of Monitor's bankruptcy. Monitor wasn't killed by any of the five forces of competitive rivalry. Ultimately, what killed Monitor was the fact that its customers were no longer willing to buy what Monitor was offering. Monitor failed to add value to customers. Eventually customers realized this and stopped paying Monitor for its services. Ergo, Monitor went bankrupt. Monitor was crushed by the single most dominant force in today's marketplace—the very force that is missing from the five forces: the customer.

■

It is therefore not really a surprise that Monitor went bankrupt. The more interesting question is: How was Monitor able to make so much money from such an illusory product for so long? Part of the answer is that Porter's claim of locating a firm's sustainable competitive advantage had massive political, social, and financial attraction for top management.

Embedded in Porter's approach to strategic planning are certain assumptions. First, strategy is about the selection of markets and products. Second, these decisions are responsible for the value that the firm creates. And third, the master decider is the CEO. "Strategy, says Porter, speaking for all the strategists, is thus 'the ultimate act of choice.' The chief strategist of an organization has to be the leader—the CEO."[18]

Strategy so conceived leads to "the division of the world of management into two classes: 'top management' and 'the rest.'" Strategy thus "defines the function of top management and distinguishes it from that of its social inferiors."[19] That which is done at the top of an organizational structure is strategic management. Everything else is the menial task of operational management.

Strategy consultants "insist on this distinction between strategic management and lower-order operational management. Strategic (i.e. top) management is a complex, reflective, and cerebral activity that involves interpreting multidimensional matrices. Operational management, by contrast, requires merely the mechanical replication of market practices in order to match market returns. It is a form of action, suitable for capable but perhaps less intelligent types."[20]

The practice of consultant-driven strategy thus fueled the mythology of the CEO as a "super-decider" and lent credence to the C-suite's soaring compensation. The attraction of this kind of strategic planning was precisely that it advanced the political, social, and financial pretensions of the C-suite.

The consultants in these strategy engagements were not usually people with deep experience or understanding of what customers might need or what. They were not generally experts in building cars or making mobile phones or building great software. They were often part-time academics or numbers-people offering financial solutions to problems that required real-world answers. That didn't matter, because their most important role was that of courtiers performing the ceremonial function of paying homage to the warrior gods in the C-suite.

Through the insistence that the C-suite alone could select the appropriate strategic plan for the future, strategic planning served the function of imposing the direction set by the top on the organization below. The fact that the plans conceived in isolation might not fit the actual marketplace, even at conception, was less important than who was deciding. Nor did it matter that in an increasingly VUCA world, even good plans selected by the top quickly got out of sync with changing conditions. The C-suite would be the ones anointed to fix such issues, with, of course, fresh consultant help.

The top-down approach to strategy was thus a corporate ritual. It was "to CEOs what ancient religions were to tribal chieftains," writes the former strategy consultant Matthew Stewart in *The Management Myth*. "The ceremonies are ultimately about the divine right of the rulers to rule—a kind of covert form of political theory." It is "like a ritual rain dance. It has no effect on the weather that follows, but those who engage in it think that it does."[21]

■

Eventually, ceremonial rain dances come to be recognized for what they are. The search for "sustainable competitive advantage" through studying the existing structure of the industry became increasingly implausible. Toward the end, even Monitor itself had abandoned trying to sell Porter's five-forces framework, though it continues to be taught in top business schools.[22]

By contrast, in Agile management, the idea that competitive advantage can be *located* has given way to the realization that competitive advantage has to be continuously *created*. Strategy is not a *place*—it's an *activity*. In Agile management, strategy recognizes that the future cannot be discovered solely by studying mountains of data based on the past or by probing what competitors might be doing. In fact, there is a recognition that in a world of rapid technological change, the future structure of the industry will almost certainly be different from the current one.

Thus, the failure of information-based strategy doesn't mean that Agile organizations don't need to pay attention to strategy. Agile management does need to be thinking proactively about longer-term opportunities and threats. As discussed in Chapter 6, the central task of strategy in Agile management is to answer the following questions: How will customers meet their needs in the future? What are the larger needs that actual or potential customers have, only part of which are addressed by current products and services? What approach can our organization take to help meet those needs in a profitable fashion? What benefits will flow from this as compared to costs? Can we do this better than any competitor? How can risk be managed?

A proper approach to strategy in Agile management means emancipating strategy from the sole province of top management and implementing a truly inclusive process in accordance with the Law of the Network. Ideas can come from anywhere, and the insights of the entire organization—and beyond—are mobilized in an effort to imagine and influence the future.

In Agile management, the myth that the top management has a lock on wisdom about the future is set aside, with the realization that strategy is innovation and innovation is everyone's business.[23] It is recognized that an organization operates as a network with a hierarchy of competence, rather than a hierarchy of authority. Multidisciplinary thinking is needed to translate potential changes into relevant opportunities for delivering fresh customer value and then implement the resulting initiatives.

Inclusion of the whole organization in strategy formulation effectively eliminates the handover hurdle between formulation and execution—a major reason for strategy failure. Devices such as hackathons and boot camps can also help by underlining the importance that the firm attaches to innovation and draw on new ideas from wherever they are.

The inspiration for these shifts is not new. It stems from Helmuth von Moltke in the nineteenth century and the military's mission command in the twentieth. Business managers need to unlearn the information-based exercises that passed for strategy in the late-twentieth-century business world.

Strategy involves leaders at all levels making decisions based on an assessment of a fluid, constantly evolving situation within an overall strategic design. It accepts that the world is probabilistic, unpredictable, disorderly, and uncertain, and that requires decentralization, spontaneity, informality, loose rein, self-discipline, initiative, and cooperation. Acceptable decisions that are made faster focus on harnessing ability at all levels.

Implementing strategy in an organizationally inclusive fashion will be impossible to accomplish in a top-down bureaucracy, with all its rigidities, layers, and unidirectional communications. Thus, rethinking business strategy for the twenty-first century means remaking the notion of management itself.

BOX 11-1

THE STRATEGY OF "DOING MORE OF THE SAME"

In 2010, strategy consultant Cesare Mainardi and Art Kleiner gave us a brilliant depiction of the phenomenon of a large-scale strategy exercise by the top management of a global company. The result? After reviewing an array of strategic options, the top management decides, yet again, to do "more of the same."[1]

It's 8 a.m. in the executive conference room of a large global packaged-foods manufacturer. . . . For the past two months, a team made up of 15 senior people has been exploring options for growth, winnowing them down to three basic strategies. Each is now summed up in a crisp 20-minute presentation. The first option focuses on innovation. . . . Under the second option, the company would get closer to its customers. . . . The third option would involve becoming a category leader . . . competing more aggressively by . . . push[ing] costs down, and completing key acquisitions.

After the screen goes blank, the CEO leans forward and asks a simple question: "Which strategy would give us the greatest right to win?" His tone, calm and direct, makes everyone sit up a little straighter. . . .

The CEO's question about the right to win sparks many levels of discussion. For several more days, spread over a few weeks, the executive team talks through its three proposed strategies in detail.

The company executives ultimately settle on the category leader strategy. It fits best with the capabilities that they already have.

In other words, the company carried out a lengthy strategic review of its options at massive expense and ultimately decided to go on doing what it had always done, with only minor modifications. The exercise was essentially a ceremony, with limited substantive content.

NOTE

1. C. Mainardi with A. Kleiner, "The Right to Win," *strategy + business*, November 2010, http://www.strategy-business.com/article/10407.

BOX 11-2

OPTIONS REASONING
AND THE PORTFOLIO APPROACH

A serious strategic planning effort will generate a number of possible options for the future. How to manage those options? A frequent problem in traditional management is that a firm has so much trouble deciding which option to pursue that it picks one and pursues that one option, willy-nilly, while ignoring the others. (See Box 11-1.)

Options reasoning and portfolio management represent a way of making investments in the future without having to back one option too soon and risk massive losses if it turns out to be the wrong one. As Columbia Business School professor Rita Gunther McGrath writes in her book *The End of Competitive Advantage*:[1]

Established organizations tend to put far too much money behind new ideas, treating them as though they know exactly what will happen, even though they are highly uncertain. One of the unfortunate consequences is that when things don't go as planned, there is an overwhelming tendency to persist, because the sunk costs look frightening to write off. This in turn often leads to painful and expensive flops, from massive product failures such as the Iridium project to disastrous acquisitions, such as the $850 million AOL basically wasted purchasing social networking site Bebo. A more effective approach in uncertain environments is to allow resources to be invested only when uncertainty is reduced, a core principle of options reasoning.

Firms will need to manage multiple potential innovations through a portfolio approach. Because the probability of the success of any particular market-creating innovation is initially low, they will need to be pursuing several innovations for each one that is an eventual success. Care needs to be taken that the portfolio includes innovations over multiple time horizons, with genuine market-creating innovations as well as efficiency or performance-enhancing innovation. (In the real world, when making decisions about market-making innovations and performance innovations, leaders obviously need to make trade-offs.)

Firms also need to avoid the risk that the portfolio becomes the bureaucratization of innovation, with a preference for low-risk, low-return, and incremental opportunities. Management must ensure cross-boundary cooperation to pursue bold, longer-term market-creating opportunities.

Equally, attention must be paid to adapting or culling businesses or programs that are no longer pulling their weight. Here top management involvement may be necessary to ensure that units don't continue to pour excessive resources into fading programs in a losing effort to keep them alive.

NOTE

1. R. G. McGrath, *The End of Competitive Advantage* (Boston: Harvard Business Review Press, 2013), 90.

EPILOGUE

12

NUCLEAR WINTERS AND GOLDEN AGES

If you will it, it is no dream.

—THEODORE HERZL[1]

S o far in this book, we have been on a journey through two different worlds.

In one world—described in the first seven chapters—we saw people and organizations in which there are healthy investments, energized workforces, continuous innovation, and strong returns. These people and organizations are at ease in a VUCA world where everyone is connecting with everything, everywhere, all the time. With the right mindset and through pursuing the Law of the Small Team, the Law of the Customer, and the Law of the Network, these organizations are progressing toward instant, intimate, frictionless value for large numbers of customers, and ultimately value for themselves and for society.

The principles they are applying to succeed in this world are simple to understand. People do best when what they do is in the service of delighting others. When they are able to work on something worthwhile with others who love doing the same thing, the group tends to get better. By working in short cycles, everyone can see the outcome of what is being done. When communications are interactive and everyone is open about what is going on, problems get solved. Innovation occurs. Customers are surprised to find that even unexpressed desires are being met. Work becomes fun. While none of the organizations we visited are flawless embodiments of these principles, they are all on a journey to get there.

In the other world—reviewed in Chapters 8 through 11—we saw people and organizations following very different principles. They are ill at ease in a VUCA world where everyone is connecting with everything, everywhere, all the time. They find themselves having to run faster and faster just to stay in place. With their disengaged workforces being told what to do, they are structurally incapable of meeting today's market requirement: continuously delivering instant, intimate, frictionless value at scale. They are mainly focused on defending the status quo and protecting their existing businesses. They are preventing themselves from moving into the future by falling into the traps of short-term shareholder value, share buybacks, cost-oriented economics, and backward-looking strategy. To hide the shortfall in real performance, there is a frequent resort to financial engineering to create a semblance of success.

In the latter group of organizations, work is deadly serious. People don't come to work with a spring in their step. Spending most of their waking hours making money for the boss doesn't bring out the best in them. In this world, some individuals and firms end up as winners, but society as a whole loses.

In terms of the relative size of these two worlds, the picture is discouraging. Today, there are many more public corporations operating in the latter mode—traditional top-down bureaucracy—than in the former mode—the emerging way of running organizations. And there are many more people at all levels of society still believing in, supporting, and promoting the traditional mode than the emerging one. This thinking, and the behavior involved in it, are reinforced by the widespread support for its assumptions and the soaring stock market that spirals ever higher.

Yet the important news is that the two worlds are on very different trajectories. The emerging world is prospering, growing, and thriving in real terms while inspiring and energizing those involved in it, and continuing to evolve and reinvent itself.

The other world is declining in real terms, and dispiriting those involved in it, even those few who are enjoying its large but ephemeral financial fruits. The true measures of its progress are not the booming stock market or the short-term benefits to shareholders but rather the sputtering real economy, the despair of large segments of the population who no longer see an improving future, the unraveling social fabric, and the political stalemate in the face of grave issues.

Even more interesting news is that within most large public corporations, there are now significant islands of practitioners of the different way of doing things and who are advocating change. Some of the islands are very large. In some global companies, there are massive divisions that explicitly see themselves as "sleeper cells" for corporate transformation.

Despite the angst with which top management sometimes views these islands of dissent, they cannot be ignored because it's simply not possible to manage modern software with traditional bureaucratic practices. In software development, there is no option but to pursue Agile management: The shift is nonnegotiable. If management won't support Agile management, the best developers go elsewhere. And since software is "eating the world" and becoming an ever more central aspect of every business model, the pressure for change from within the organization is never-ending.

The passion and ambition with which these islands advocate change can sometimes set traditional managers' teeth on edge. The issues are partly social: It's as if the servants have forgotten their place and are taking over the chateau. The resulting culture clash is amusingly described by Paul Ford in a *Bloomberg Businessweek* article. It begins with the quandary of a previously successful executive whose experience and skills are useless in coping with languages he doesn't understand, management practices he cannot grasp, people he doesn't feel comfortable with, and threats to his survival as a manager that are all too real. Software development is consuming an ever-larger part of his budget while it is becoming ever central to his, and his organization's, future. Although the article depicts a happy ending as the manager gradually grasps what is going on, in real life the outcome is often otherwise.[2]

Yet over time, as top executives themselves are coming to recognize the need for the organization to up its own game to cope with the fast-changing marketplace, the pull for Agile management increasingly comes from the top. In effect, the question shifts from "Why do we have to change?" to "Why can't we have what they're having?"

Leading management journals such as *Harvard Business Review, The Economist,* and the *Financial Times* have also taken up the cause.[3] Boards and asset managers are showing an increasing interest in the genuine health of the companies they invest in—not just the companies' share price.[4] As large parts of the business ecosystem heads off the rails, business is increasingly recognizing the need to act out of enlightened self-interest. Business leaders are thus not simply practitioners of capitalism. They are its stewards, with a responsibility to enhance the sustainability of the market system.

Why is the transition taking so long? Why is there such a large gulf between these two worlds? Why, at a time of unparalleled technological innovation, are so many organizations protecting the status quo and propping themselves up with financial engineering, rather than promoting real growth with rapid transformation? We have already seen some of the reasons, including the deep nature of the change, the challenge to existing habits, attitudes, and practices, and the massive financial incentives to the C-suite to maintain the status quo.

From a historical perspective, it's no surprise that these two different worlds exist side-by-side for extended periods. After all, this is not the first time in history that the world is encountering the emergence of "new technology," resulting in a "new economy" that requires "a new kind of management" and that is accompanied by "a financial frenzy," while many firms are in denial and stick with "more of the same." Professor Carlota Perez's book *Technological Revolutions and Financial Capital: The Dynamics of Bubbles and Golden Ages* gives us a brilliant description of similar phenomena occurring in earlier historical periods that shed light on our current situation and the potential outcomes that lie ahead.[5]

"History doesn't repeat itself," as Mark Twain allegedly quipped, "but it does rhyme."[6] The specific challenges of our own age are of course unique to the current period—globalization, deregulation, knowledge work, and astonishing new technology, particularly the Internet and the growing role of software in all businesses. While in these respects, our current situation is unprecedented, history has lessons for all those interested in accelerating the transition to a better way of running organizations, whether they be Agile managers, business leaders, or simply concerned citizens.

Thus, over several hundred years, we have seen recurring patterns. As explained in Box 12-1, clusters of innovations came together to create technological revolutions. There were successive transitions in the 1790s (canals), the 1840s (rail), the 1890s (steel), the 1920s (mass production), and the 1990s and beyond (computers and communications). In each case, there were astonishing new profit possibilities for those with eyes to see and money to invest. In the process, new ways of running organizations were developed to cope with the new technology. These changes in turn gave rise to successive waves of financial frenzy, with an expectation of high returns and an overinvestment in the new profit possibilities. In each case, the prospect of vast new fortunes was too tempting to resist.

Hot money flowed in. Investors not only exploited the real opportunities created by the new technology but also set out to make money out of money when investment in the real economy proved scarce. The financial sector then took over the economy, as firms lost interest in the real economy of goods and services, favoring the casino economy. For a period, the extraordinary returns of the overheated "new economy" and "making money from money" were the central preoccupation. Everyone thought that the gains would continue.

The winners in an economy based on making money out of money got richer while average citizens saw few benefits. In the financial frenzy, investors pursued chimeras. Eventually a big financial crash—or series of crashes—occurred and large parts of the economy were laid waste. In the more productive bubbles, genuinely better ways of organizing, managing, and consuming had been put in place. Investments had been made in new technology and infrastructure. New kinds of organizations had been created and new ways of running them had been discovered.

When the crash or crashes were severe enough, and the national leaders were wise enough, new institutional arrangements were put in place to stabilize the economy and bring the rampaging financial sector under control. The provision of real goods and services for real people was put back at the center of the economy, while the financial sector and the preoccupation with making money out of money were assigned supporting roles. When this happened, a kind of "golden age" could follow, with moderate but steady economic growth and widely shared benefits. Broad segments of the population prospered. The economy was back in a virtuous circle.

But when the crash was not severe enough to awaken society from its illusions of exponential returns from making money out of money, or when society's leaders were not wise enough to put in place the institutional changes needed to rein in the rampaging financial sector, then various kinds of national decline ensued. Instead of a "golden age," there was a "nuclear winter."

In some cases, the casino economy rolled on and the financial sector continued to rampage, with a deterioration of the real economy, persistent financial crashes, and worsening income inequality. In other cases, there was an overreaction to the excesses of the financial sector with excessive regulation that prevented the healthy functioning of the banks, thus crippling the future evolution of the real economy. In still other cases, the events led to a state-run economy with disastrous political consequences.[7]

■

Where are we today? The technology bubble of the 1990s was driven by a kind of opportunity pull, whereas the bubble of 2001 – 2007 was driven by the push of easy credit. "In the first case," writes Perez, "it was the excitement about new technology that attracted the money into the casino, almost regardless of cost; in the second it was the excitement about abundant easy money that pushed investors to seek new ways of making money out of money."[8]

Today, following the government bailout of the banks after the crash of 2008, the casino economy rolls on. After the meltdown, massive efforts were made to clean up the banks and put in place regulations aimed at restoring trust and confidence in the financial system. But the net result in terms of dealing with root causes has been mixed.[9] The financial sector is still oversized and the financial regulations create excessive bureaucracy for the banks without providing effective protections against wrongdoing. There is now talk of rolling back the protections that were put in place. In effect, society is at risk of perpetuating the egregious behavior that brought on the crash of 2008.

The current boom in both corporate profits and the stock market at a time of lackluster economic growth reflects a combination of opportunity pull (the Internet of Things), and the push from the continuing abundance of easy money. As of mid-2017, investors are salivating at the prospect of tax relief and the possibility of bringing back several trillion dollars held overseas by corporate America and transferring that to shareholders through share buybacks. Many public corporations are caught in the traps of short-term shareholder value, cost-oriented economics, and backward-looking strategy. There are many exceptions, and in most cases, there is agitation for change from within. But the prospect of continuing economic stagnation, growing income inequality, further financial crashes, and a prolonged "nuclear winter" still looms.[10]

Getting to a new "golden age" for society, with steady economic growth and widespread prosperity, will require strong leadership from many actors across society. (See Box 12-2.) A large part of the challenge must be met by the private sector rededicating itself to creating fresh value to customers with real goods and services, thus contributing to the growth of the economy, rather than financial engineering, extracting value, and making money out of money.

One object of this book has been to draw attention to the body of knowledge that is available to those involved in running organizations and who want to do so more productively. Our understanding of how to do this is already substantial and is reflected in the practice of tens of thousands of organizations around the world. We now have deep knowledge about what is involved. We know what to do differently.[11]

In effect, a new way of getting work done and running organizations has emerged. It's not just a big idea—it's a transformative idea.

It is different in kind, scale, and impact from big ideas that are aimed at improving the world, such as a robot that can outperform students or an ultraviolet light that can kill superbugs.[12] These important ideas form the fodder of conferences and TED talks. They are important ideas, but, even when embraced, they don't fundamentally change the way we think about the world. They are added to and woven into our existing way of life without basic change. They don't upend our worldview. We embrace them within our existing frame of reference.

Transformative ideas are ideas that change our frame of reference. They can't be evaluated *within* our existing frame of reference because they create a *new* way of understanding and interacting with the world. They lead on to ever more productive hypotheses as to how to improve the future.

We already discussed in Chapter 3 an example of a transformative idea—the Copernican revolution in astronomy. It wasn't just a big idea in astronomy: It led to a different way of thinking about the universe and was accompanied by vast indirect social and political consequences. It became an evolving frame within which all other ideas in astronomy were evaluated, along with changes in the way society is organized. It didn't appear fully articulated by Copernicus on day one. It steadily grew with the contributions of Galileo, Kepler, Newton, and eventually Einstein. Over centuries, it became an ever-larger idea.

Similarly, the Copernican revolution in management—call it by whatever name you like—isn't just one more competing idea for improving the way organizations are run, like a new process for teams or a better kind of brand or talent management. It is a fundamentally different way of thinking about how human beings collaborate to get big things done in the world and hence a new frame for society in evaluating all other ideas for accomplishing that.

This different way of doing things generates a prospect that is genuinely exciting. It helps understand the present and points to a better future. It is a practical path toward a world that is better for those doing the work, better for those for whom the work is being done, better for the organizations that coordinate the work, and better for society.

The Copernican revolution in management offers both a different vision for the future and a multidimensional critique of how things are currently done. Thus, it presents a *practical* critique of much of current management, which is unable to keep pace with a fast-changing world in which the customer has become central. Among other things, the new way of doing things is essential to managing modern software. As all firms become dependent on software for success, Agile management will inexorably make its presence felt across the whole organization.[13]

Agile management offers a *financial* critique of current management. It shows that even on its own terms a sole focus on enhancing shareholder value doesn't work. Shareholder value thinking is a trap that inexorably leads to a preoccupation with the short term and ultimately destroys shareholder value. The new way of doing things doesn't just lament these flaws but offers an implementable way of resolving them.

It offers a *legal* critique of today's management. The deployment of gargantuan resources—some $7 trillion over the last decade—in share buybacks is effectively stock-price manipulation and must be ended to restore the financial integrity of the economy.

It presents an *economic* critique of the way organizations are currently being run and points to the inevitable conclusion that an economy based on extracting value from corporations cannot end well. Agile management challenges economists to recognize that the determining element of much of modern economics is now value more than cost.[14] It implies setting aside the mythology of "the invisible hand" of the marketplace and examining what is actually happening. As the Nobel Prize–winning economist Joe Stiglitz has pointed out, "The reason that the invisible hand often seems invisible is that it is often not there."[15]

It offers a *moral* critique of society's current leaders. The institutionalized self-interest embodied in extracting value from corporations for shareholders and their executives through share buybacks is unethical. Current incentives lead executives to do what they should know is wrong. When leaders of a society routinely act in ways that are unethical, they undermine and ultimately destroy the values of society.

It offers a *political* critique of a society that accepts boring work, stock-price manipulation, and growing income inequality as inevitable and acceptable norms. The new way of running things implies a vision of a society that is organized and run differently. It is political in a broader sense than the platform of any specific political party. It is neither Democratic nor Republican. It is conservative in the sense that it flows from the best that has gone before. It builds on individual freedom and self-management and entrepreneurialism and innovation and independence from government intervention. It is democratic in the sense that it is egalitarian in spirit. It aspires to create prosperity for all rather than extract value for the few. It offers a view of society in which corporate raiders and the banks play a supporting role rather than a dominant, determining force in our society.

It offers a *philosophical* critique of a society that confuses goals and results. Shareholder value is a result, not a goal. When a measure or a result becomes a goal, it becomes corrupted as a measure. This confusion is a root cause of many of our current economic problems.

Thus, the new way of getting things done is a transformative idea. Like all transformative ideas, it creates an intellectual frame by which other ideas can be evaluated. It offers a way out of practical, financial, economic, social, political, and ethical dilemmas of our time. It establishes criteria by which progress can be measured and goals are set. It creates a framework within which politicians and political debate can operate. Instead of offering a jigsaw puzzle of little fixes, it provides a broad theory of how organizations—and society—can function for the better. It is a coherent story that offers the prospect of a new age.

If we are right in thinking that a new age is emerging, when did it begin? Was it Helmuth von Moltke's *Auftragstaktik* in the 1860s? Was it the Toyota Production System in the 1960s, which later became known as lean production? Was it the Agile Manifesto of 2001? Was it 2016, when the citadel of general management, *Harvard Business Review*, embraced Agile?[16] Or is it an age that is still struggling to be born, one that will only be fully alive at some time in the future when more of its components and details are fully worked out? The inquiry is as practically pointless as asking, When exactly did the Copernican revolution in astronomy begin? Was it with Aristarchus of Samos? Or Copernicus? Or Galileo? Or Kepler? Or Newton? As with any transformative idea, the age of Agile has historical roots going back a long way. It will take many more decades for its final fulfilment.

The fact that a new age is emerging now is not an accident. We live in a time when many sense, like David Brooks, that "the very nature of society is up for grabs." [17] Some grand ideas of the past—Christianity, the Enlightenment, Marxism—have declining influence. Other ideas, values, and assumptions that prior generations accepted without question, such as democracy, truth, science, freedom of the press, and morality, are also under siege. Fundamental assumptions are being flouted by populist political leaders on a daily basis. A vision of a future dystopia in which the rich get richer while the bulk of the population trudges along, doing boring work with stagnant or declining compensation, or even no work at all, is an increasingly worrisome prospect. There is a growing concern that the current social and political order could even come unstuck.

The emerging age is a response to the feeling that there *must* be a more energizing, prosperous, and meaningful mode of living and working for the entire population, which can be imagined, experimented with, and implemented.

The emerging age offers a promising path forward. It's the work of many minds, hearts, and hands. It's incomplete in many respects.

For the most part, it is being carried out by action-oriented men and women, not by academics or policy wonks. Until recently, it has been little written about in academia, or even in general management journals, with the notable exception of *Strategy & Leadership*.[18] Nevertheless, it is now happening on a large scale and moving forward relentlessly, driven by heroic actors in the workplace who have rolled up their sleeves and set out to get things done in a better way, rather than just talking about them. These people are real-world pioneers. While it's good to talk and write about policy issues, it's even better to have people who do something about them and put their thinking into practice. This is policy and politics at the most real and essential level.[19]

If the emerging age is to prevail, it will require seeing the truth squarely and being fully committed to its values. The tasks involved will include breaking ranks when necessary, facing up to unpleasant truths, and having the courage to go beyond what has been tried and found wanting. If leaders of the movement stray from core values, they will also need to be called out and held to account.

As support for the new way of doing things becomes steadily more visible, debates will continue about the details of implementation, emphasis, terminology, and so on. Improved practices will be discovered. More sophisticated measures will emerge. Wrong turnings will be recognized as such. Some practices will be discarded. But we don't need to wait for these course corrections in order to move ahead. The broad direction forward is now clear.

The emerging age thus offers the possibility of a great awakening—a foreshadowing of a transformation in the way our organizations—and our society—function. The ideas in this book constitute an invitation to recognize, applaud, encourage, nurture, and disseminate the possibility and protect it from those entrenched interests that are intent on preventing it from happening.

While the new age represents a vast social agenda, one of the exciting things about it is a preoccupation with keeping things small. Unlike the Gothic cathedrals of the past or the skyscrapers of today's global conglomerates that seek to intimidate us with their size, the age of Agile is more like early Renaissance architecture, which did its utmost to keep things on a human scale.[20]

When managers first encounter this way of doing things, their reaction is often to wonder why it is so small. With such small teams, how could it possibly solve huge global problems and operate at scale? Happily, the new way of organizing enables networks of small teams to operate at scale without sclerosis.

In the emerging age of Agile, the people doing the work have a clear line of sight to the people for whom the work is being done. It makes individuals conscious of their powers as complete, thinking, responsible human beings, liberated from internal and external contradictions.

In the twentieth century, the phrase "the dignity of man" died on people's lips as a travesty as they watched the horrors of industrial-scale warfare, the daily crushing of the human spirit in soulless workplaces riddled with contradictions, and the relentless pursuit of self-interest by the well-to-do.

By contrast, in the emerging age of Agile, the dynamic focuses on human beings creating delight for other human beings. When an organization—or a society—is populated by people with this mindset, it can be at one with itself, at one with those for whom the work is being done, at one with those who are doing the work, and at one with the wider society in which it operates. In such a world, the meaning of "the dignity of man"—and women, too—is fresh and invigorating.

BOX 12-1

THE HISTORY OF GOLDEN AGES
AND NUCLEAR WINTERS

From a longer-term historical perspective, business that is analyzed in terms of quarters, years, or even decades may miss the longer cycles of change that are under way. Recalling this history can shed light on our present dilemmas as well as possible next steps. It can be helpful to simplify the history into five main eras:

♦ The 1790: canals
♦ The 1840s: rail
♦ 1880s to 1890s: steel
♦ 1910s to 1920s: mass production
♦ 1970s and beyond: computers and communications

The 1790s: Canals

Few people today know that in the 1790s in England, the exciting new technology was, of all things, canals. With the onset of the Industrial Revolution, as mechanized factories began to transform the rural English economy, there was a rapid expansion of roads, bridges, ports, and canals to support the growing flow of trade. A phenomenon known as Canal Mania emerged, as money flowed in to fund more canals. Some of the hot money was seeking refuge from the ongoing French Revolution. The investments funded canal after canal, including those that weren't needed. Great fortunes were made until the Canal Panic of 1797, when it was suddenly realized that there had been an overinvestment in the new technology and the financial bubble burst. A great deal of the money that appeared to have been made evaporated. There was a lot of waste and pain, but in the end, the Canal Mania had also funded infrastructure that remade the British economy.

The 1840s: Rail

In the 1840s, another hot new technology exploded onto the English scene: railways. Sparked by the transformational impact on society of rail travel, there was an amazing investment boom. Everyone wanted to invest in this exciting new technology, which offered the prospect of immense new wealth. Money poured in from all over, resulting in the Rail Mania of the 1840s. Huge new investments were made and great fortunes were created, until the Rail Panic of 1847. In the confusion, some people became very rich while others went bankrupt. The poor were left behind. When the dust settled, England had built far more railways than it could possibly use.

There followed a calmer period in which the overbuilt rail network was rationalized. Here, too, the Rail Mania involved a lot of waste and pain but it was not all bad: It funded infrastructure that had once again remade the English economy. During this period of relative economic stability, the new management practices instigated by the arrival of railways were institutionalized and some sharing of benefits more widely across the population was achieved. For a period, a kind of "golden age" ensued.

1880s to 1890s: Steel

Later, as the English economy became bogged down in increasing financialization and overseas investments, the economies of the United States and Germany moved to the fore. In the 1880s and 1890s, the hot new technology was steel. A huge transformation of the world economy was under way with transcontinental trade and travel, accompanied by international telegraph and electricity. Again, there was excited talk of "a new economy" with "new technology" and "new financial possibilities." Financial markets received a massive infusion of cash. Everyone wanted to be in on it. That is, until a series of crashes came in various forms in the United States, France, Italy, Australia, New Zealand, South Africa, and Argentina.

Chastened by the crashes, the financial sector in the United States and Germany returned its focus for a while to funding the real economy, which stabilized the situation for a period. The result was less happy for Argentina, which ceased to be a major player in the world economy. Great Britain, which hadn't invested in steel, also began losing its leadership position in the global economy. Once again, the change involved a great deal of waste and pain, but it had funded infrastructure for the new global economy. Thereafter, there was a period of relative calm in which excessively powerful trusts were broken up and income inequality was addressed.

1910s to 1920s: Mass Production

In the early years of the twentieth century, investors once again became excited by the prospects of mass production with Henry Ford's auto industry. It offered once again the potential to transform society. Investors emerged to take advantage of the opportunities. By the 1920s, the stock market had grown to become the primary engine driving the American economy. Stocks were guaranteed to grow in an unending bull market. Everyone wanted to cash in. That is, until the collapse came in 1929.

The ensuing recession was deep and prolonged, until the financial sector, through legislation like the Glass-Steagall Act, was reconnected with the "steady" real economy of firms producing real goods and services for real customers. It took a while for the reconnection to take effect. When it did (and in parallel with the heavy investments necessitated by the Second World War), there followed several decades of an economic "golden age" in the United States with high economic growth and shared prosperity.

1990s and Beyond: Computers and Communications

In the final decades of the twentieth century, hot new technologies were emerging: all-pervasive computing and digital communications. Computer chips were powerful and cheap. They opened innumerable business possibilities. Around the world, they transformed the way people lived and worked. In the decades that followed, huge fortunes were made and lost as part of the transformation. The proof that a "new economy" had arrived was found in the good times of the prosperous 1990s. New profit possibilities appeared at every turn. Making money became a subject of universal interest as everyone rushed to take advantage of the new investment opportunities.

Emboldened by the amazing gains that were possible from the new technology, the financial sector's investments went far beyond technology. The returns were amazing, but again, they were unsustainable. The dot-com bubble burst in 2000, creating the equivalent of a "nuclear winter" over Silicon Valley.

For some, the party was over. The "wizards" of the dot-com era in Silicon Valley were forced to sober up after an era of "irrational exuberance." Computers were transforming society, but it turned out that the "new economy" was governed by some fundamentals of the old economy: Value to real customers and earned profits still mattered.

Yet despite the pain and waste caused by the bursting of the dot-com bubble, when it was over, valuable physical and institutional infrastructure for the new economy of computers and telecommunications had been put in place. Massive amounts of fiber optic cable had been laid. Firms had modernized their computer systems. In Silicon Valley, a vast social network had been built that could foster the next generation of economic players like Apple, Amazon, Facebook, and Google, in which the new ways of running organizations would flourish. The process had been difficult, but in the end, some of the productive institutions of society had been remade.

By 2001, investment in Silicon Valley shriveled. But in the financial sector, the appetite for amazing gains from great risk-taking remained unabated. During the dot-com frenzy, the financial sector had steadily lost interest in the real economy. The financial sector was no longer satisfied with the returns that came from financing the production of goods and services for real customers. The financial sector sought new ways to get exceptional returns by making money out of money.

Within a few years, with the indulgence of the central bank, the financial sector was once again creating amazing gains—this time, from real estate. For a brief period, America was once again celebrating. The economy was booming. Everyone who owned assets was getting wealthier, even though the warning signs were everywhere: too much borrowing, foolish investments, greedy banks, exotic new financial instruments that were deliberately designed to be opaque, regulators asleep at the wheel, politicians eager to promote homeownership for those who couldn't afford it. Distinguished analysts who openly predicted that this would end badly were ignored.[1] When Lehman Brothers fell, the financial system froze and the whole global economy almost collapsed.

Wall Street was able to avoid the nuclear winter that had afflicted Silicon Valley, with the help of a government bailout of the big banks. Main Street was not so lucky. Large numbers of small and medium enterprises went bankrupt. Jobs were lost. Savings were destroyed. Real property values plunged. Houses went underwater and mortgages were foreclosed. Median incomes declined.

A large stock of unneeded housing had been built, but it was largely unproductive investment. Unlike earlier financial bubbles—canals, rail, steel, mass production, and the dot-coms—investments in housing that people couldn't afford left the economy in no better position to move forward or compete internationally. The housing that had been built was pure consumption. The housing bubble had few positive elements, except personal benefits for the financial wizards. For some years, the stock of unused housing sat like a dead weight on the economy, holding it back.

As of mid-2017, the soaring stock market at a time of meager economic growth reflects a combination of opportunity pull (the Internet of Things), and the push from the continuing abundance of easy money. Many public corporations are caught in the traps of short-term shareholder value, cost-oriented economics, and backward-looking strategy, while more forward-looking firms are advancing the age of Agile.

NOTES

1. S. Denning, "Lest We Forget: Why We Had a Financial Crisis," *Forbes.com*, November 22, 2011, https://www.forbes.com/sites/stevedenning/2011/11/22/5086/.

BOX 12-2

HOW THE CHANGE MIGHT HAPPEN:
AN AGENDA FOR ACTION

The goal of Agile management—delighting customers in ways that are financially sustainable—is at odds with thinking that is still pervasive in public corporations—namely, that the goal of a corporation is to maximize shareholder value as reflected in the current stock price. Shareholder value thinking is taught in business schools, presumed in daily financial news reporting, accepted as a go-to for any executive of a large public company, adopted as the modus operandi of activist hedge funds, endorsed by regulators, practiced by institutional investors, accepted by citizens in their retirement planning, and embraced by analysts and politicians.

As a result, leaders at all levels of a public corporation find themselves subject to enormous pressures to focus on short-term profits and increasing the current stock price for the benefit of shareholders, at the expense of other stakeholders; in the process, they undermine the goal and practice of Agile management. We are dealing here not so much with iniquitous individuals, but rather a whole system that has gone awry. Individuals find themselves trapped in a system they cannot control.

Given the negative consequences of this system, society faces a set of significant financial, social, and political issues. We now know that shareholder value thinking generally has the opposite effect of what was intended. We know that it generates socially unproductive short-termism, cripples investment, and systematically destroys value. We know that a total focus on the extraction of value from corporations for the benefit of their shareholders is harmful to those corporations, unproductive for customers, prejudicial to their employees, and ultimately harmful for society itself. We know that missteps in the financial sector have helped undermine the centuries-long tradition of bankers as the pillars of the society. We can see that these missteps have led to the unfair demonization of all businessmen and bankers. We can also see that such demonization causes citizens to lose sight of the beneficent role of the business as the engine of national prosperity and a healthy financial system as a key ingredient of the good society.[1]

Given that we know how to run organizations in a better way, what would be involved in enabling business and banking to head in a better direction? Many different actors will need to play roles, including:

- CEOs
- Boards of directors
- CFOs
- Financial sector
- Regulators
- Rating agencies
- Investors
- Politicians
- Business schools
- The media
- Thought leaders

Let's look at each of the roles.

CEOs

The first step to achieve change concerns the behavior of CEOs. As noted in the first seven chapters of this book, while many CEOs are still managing quarter to quarter, looking at short-term results in a traditional paradigm, cutting costs, and extracting value for shareholders, other CEOs have built a culture of innovation and long-term value by developing their people and declining to participate in the charade of offering quarterly guidance on the future of their earnings.

There are now public firms—like Apple, Amazon, Google, and Unilever—along with many privately owned firms and small and medium enterprises that have recognized the primacy of delivering value to customers. None of these firms are perfect, and some, like Apple, have on occasion engaged in large-scale share buybacks. But by and large, these firms are focused on adding long-term value to customers and the corporation, rather than maximizing the stock price in the short term.

Such firms are becoming a steadily larger part of the economy, as this radically different management focus on customers ahead of share price is the basis for enduring success in a marketplace in which the customer is collectively in charge.

Yet CEOs of publicly owned firms still live in a world in which there are strong pressures to do the opposite. When a whole society is on the wrong track, as Hannah Arendt has pointed out, many will follow unthinkingly and do whatever the incentives dictate.[2]

CEOs pursuing shareholder primacy are thus not wrongdoers in the sense of people who deliberately set out to do wrong. They are people who find themselves in situations where actions that are causing great harm to their firm and to society are seen as normal and expected and even strongly rewarded. Many CEOs take the easy path and simply accept the status quo, waiting until society as a whole changes. Yet CEOs are obviously capable of doing some hard thinking and understanding how their actions impact other people. They need to think through how they can act more like stewards of management, not merely practitioners of the status quo. As outlined in the following sections, actions by society as a whole can help spur CEO thinking and speed up the pace of change.

Boards of Directors

We cannot expect most CEOs to act differently when compensation committees dangle multimillion-dollar bonuses in front of them to encourage them to act irresponsibly. Stock-based pay itself is a large part of the problem and needs to be reined in. By its very nature, it encourages the C-suite to focus on the short term. "Overall the use of stock-based pay should be severely limited," says professor Bill Lazonick. "Incentive compensation should be subject to performance criteria that reflect investment in innovative capabilities, not stock performance."[3]

The problem isn't individuals. Structural solutions are needed. The composition of boards and compensation committees needs to be revisited. "Boards are currently dominated by other CEOs, who have a strong bias toward ratifying higher pay packages for their peers." Other risk-takers such as "taxpayers and workers should have seats on boards," says Lazonick. "Their representatives would have the insights and incentives to ensure that executives allocate resources to investments in capabilities most likely to generate innovations and value."[4] Executives should be rewarded for progressive improvement in delivering value to customers. Investors should be given detailed information on how the C-suite is performing in that regard, not merely on short-term profits.

Chief Financial Officers

Another key set of actors are the CFOs. They often act as guardians of "the single objective financial function" by which shareholder value theory is enforced in decisions taken daily throughout the whole organization. Every decision and action is thus evaluated in terms of its impact on short-term earnings per share, not on whether it is delivering value to customers.

The financial function needs to be redefined so that it takes into account short- and long-term interests of the organization, particularly the primary role of customers. Financial considerations are still present, but they are not the sole driver of actions. As we saw in Chapter 10, different metrics need to be used and deployed with a different mindset.

The new financial function implies a new role for the CFO. Instead of CFOs being single-mindedly focused on reducing costs, they must serve in a supporting role of ensuring that the firm continues to be profitable as it steadily innovates and adds value to customers. Profits become a result, not a goal.

The shift is not just about the adoption of new metrics and applying them with a different management mindset. It's also about a shift in power. It's about who is calling the shots in the firm. Currently, the finance function is in charge because it can quickly move the needle on quarterly profits by cutting costs and thus ensure the C-suite's bonus. CFOs have not always known or cared much about the firm's products and services. People who understand the customers and create real value for them—the product engineers, the designers, and the marketers—have been shoved aside because they couldn't "move the needle" in terms of quarterly profits. In the emerging age of Agile, those sidelined players must now have a louder voice.

Needless to say, CFOs are not going to abdicate their position of power easily, voluntarily, and unilaterally. Boards and CEOs must establish new rules of the game in which customer value becomes the guiding principle of corporate action, not shareholder value, quarterly profits, and cost-cutting preoccupations.

The Financial Sector

In a healthy economy, the financial sector, particularly banks, plays a crucial role. The financial sector acts like a circulatory system, with money flowing like blood to preserve and strengthen the health of the economic body. It translates products and services into exchangeable financial instruments that facilitate trade in the real economy. Through deposits, citizens' savings are channeled to businesses that can use them productively. Through mortgages, workers can trade their promise of future wages for a home. Through insurance, homeowners are able to share financial risks and avoid financial catastrophe. The financial sector enables jobs to be created, shops to be built, houses to be bought, investments in the factories and facilities to be made, and risks to be offset.

When banks act like this, bankers are seen as stellar citizens and pillars of the community. This virtuous circle is what generally happened in the U.S. financial sector for half a century after the excesses of the 1920s. The subsequent excessive financialization and the phenomenon of money chasing money have contributed to a state of secular economic stagnation and an image of Wall Street as a bunch of smug, bonus-hungry confidence-men in pinstripe suits. Could this change?

"Financial capitalism is an invention," writes Robert Shiller, "and the process of inventing it is hardly over. The system has to be thoughtfully guided into the future. Most importantly, it has to be further expanded and democratized and humanized, so that we may reach a time when financial institutions will be even more pervasive and positive in their impact. That means giving people the ability to participate in the financial system as equals, with full access to information and with the resources, both human and electronic, to make active and intelligent use of their opportunities. It will mean that they truly consider themselves part of modern financial capitalism, and not the victims of the aggressive and selfish acts of a cynical financial establishment."[5]

Enlightened self-interest requires that the financial sector curb the current excesses and once again resume its role as the circulatory system of the real economy. Regulations, while both necessary and helpful, have potential for playing a supporting role in the needed transformation. But the transformation must be led from within the financial community itself.

Regulators

Regulators must also play a central role. An obvious step is the repeal of the "safe harbor" provided by Rule 10b-18 of the Securities Exchange Act, which in practice allows corporations to engage in stock-price manipulation with impunity.

More important, regulators must shift their view of the function of regulation beyond the identification of wrongdoing by individuals—usually individuals below the level of the C-suite—and take on the more important function of identifying systemic failure. When most big corporations are involved in value-destroying activities on a routine basis, then the challenge for regulators is not one of responding reactively to the odd individual wrongdoer. Regulators must think proactively as to how to inspire change in a marketplace where much of the prevailing way of doing business is at odds with the interests of society.

For instance, regulators should require regular reporting on progress toward the organization's primary goal: innovation and delivering value to customers. At present, most big corporations give investors little information on this subject. Yet a standard methodology for measuring customer satisfaction is now available. Hundreds of major firms are now conducting audited Net Promoter Scores and some firms are publishing their results, such as Philips, Schwab, Intuit, Progressive, and Allianz. Reporting such information, duly audited, to shareholders should become compulsory.

Rating Agencies

Rating agencies must also play an important role. They were complicit in condoning and even rewarding some of the riskiest practices in the 2008 meltdown and they received significant compensation for doing so. They must take a harder look at the noxious consequences of share buybacks, particularly when funded by anomalously cheap borrowing.

Investors

The role of investors is also central. Traders, including high-speed traders who are only focused on short-term fluctuations in asset prices, must be more tightly regulated. International agreements need to be developed to impose taxation on short-term trades, which are a principal cause of market volatility.[6]

Some investors will continue to chase short-term returns. If open-market share buybacks are outlawed and short-term trades are subject to taxation, value extraction will have to come in the form of publicly announced dividends and will be less easy to justify. Firms will be encouraged to earn their returns, rather than simply manipulating the stock price.

Investors interested in longer-term shareholder value must learn that chasing short-term gains is a fool's errand. As long as they focus on, and reward, short-term gains in quarterly earnings, without regard to the fundamentals of how those gains were generated, they are not only engaging in predictably irrational behavior—they are also reinforcing conduct that will soon undermine those very returns. They need to shift their attention to those aspects of corporate performance that create real value for shareholders (i.e., customer delight).

Institutional investors have a particular responsibility in showing the way forward and refraining from participating in behavior dictated by short-term moves in the share price. In an open letter to corporate

America in 2015, Laurence Fink, the chairman and CEO of BlackRock, the world's largest asset manager, called on companies to stop borrowing to boost dividends and increase share buybacks. Yet institutional investors also need to deal with the fact that their own managers are often rewarded on the basis of short-term returns. Thus major shareholders themselves must cease being complicit in the extraction of short-term value.

Politicians

Politicians must also play a role. We need political leaders who are willing to stand up and fight for what is right. Is this conceivable in our fractious political climate, when corporate lobbyists have a disproportionate influence on politics? Doris Kearns Goodwin in her book *The Bully Pulpit* (Simon & Schuster, 2013) discusses the massive economic problems in the United States in the early-twentieth century and tells how President Theodore Roosevelt had only his voice to use against the injustices going on in banking, labor, and industry. Congress didn't want to hear or do anything about it. The key was the stories he told about real people. The stories were repeated everywhere. Several decades later, President Franklin D. Roosevelt also fought to achieve a fairer society with responsible business leaders. So, it was bold leadership that built up the momentum for reform that eventually succeeded. Today, we need political leaders who have the understanding and courage to make the case for change.

Business Schools

Academics also have a responsibility to stop embracing shareholder value theory without question and start systematically teaching the better idea: that the primary purpose of the corporation is to create a customer. Economic textbooks that assume the validity of shareholder value theory must be rewritten.[7]

The Media

Analysts and the press also have a responsibility to upgrade their thinking about what is good performance for a corporation. When analysts exult because a stock is "on fire," they need to be analyzing what the underlying basis of its performance is. Is the stock advancing because of genuine strong performance, or is it financial engineering of the worst kind? Many analysts have already spoken out, as shown in the many articles cited in this book. This thinking needs to become the norm.

Analysts and journalists should also point out good behavior. They should draw attention and celebrate firms that are excelling in delivering customer delight, instead of fixating over minor shifts in quarterly profits.

Thought Leaders

Global institutions like the Drucker Forum have a historic responsibility to bring together thought leaders and alliances that are already active in this area, such as the Coalition for Inclusive Capitalism, the B Team, Conscious Capitalism, the Skoll Foundation, and the Kauffman Foundation. They should not only forge agreement on the root causes of our current economic and management dysfunctions but also form an active global coalition that works to achieve a better way forward.

Is such a vast social and political agenda realistic, given that most of the power, resources, and incentives currently rest with those defending the status quo? This is the natural condition of every vast social movement, such as the advance of democracy, the elimination of slavery, the repair of the hole in the ozone layer, and the halting of climate change. Vast social movements always confront daunting challenges. They can, however, take heart from the dictum, sometimes attributed to Margaret Mead: "Never doubt that a small group of thoughtful, committed citizens can change the world. Indeed, it is the only thing that ever has."

NOTES

1. R. Shiller, *Finance and the Good Society* (Princeton, NJ: Princeton University Press, 2012).
2. H. Arendt, *Eichmann in Jerusalem: A Report on the Banality of Evil* (New York: Viking Press, 1963).
3. W. Lazonick, "How Stock Buybacks Make Americans Vulnerable to Globalization," Working Paper, East-West Center Workshop on Mega-Regionalism: New Challenges for Trade and Innovation, March 11, 2016, https://papers.ssrn.com/sol3/papers .cfm?abstract_id=2745387.
4. Ibid.

5. Shiller, *Finance and the Good Society*, vii–viii.

6. D. Clliggott, "How to Avoid Another Market Crash," *Fortune,* September 2, 2015, http://fortune.com/2015/09/02/stock-market-volatility-china/.

7. S. Denning, "How Modern Economics Is Built on the World's Dumbest Idea," *Forbes.com*, July 22, 2013, https://www.forbes.com/sites/stevedenning/2013/07/22/how-modern-economics-is-built-on-the-worlds-dumbest-idea/.

ACKNOWLEDGMENTS

It is impossible to thank individually the huge number of people who have contributed to the creation of this book. All I can do is to signal here a few who have been particularly helpful.

This book obviously draws on the massive literature of management, leadership, as well as Agile and lean. I have done my best to indicate in the text itself the sources of my thinking so that readers can immerse themselves more deeply in these vast streams of thought and practice.

I am indebted to a small group of collaborators that meets weekly to share their experiences of the emerging workplace, including Rod Collins, Jay Goldstein, Andrew Holm, Dawna Jones, Thomas Juli, Nancy Van Schooenderwoert, and Peter Stevens.

I would like to acknowledge the help of my collaborators at the SD Learning Consortium, who continue to energize and inspire me, including Vanessa Gamboa Adams, Matt Anderson, Lindsay Bennett, Aaron Bjork, Chris Connors, Susan Gordona, Kevin Grady, Chad Lindbloom, Paul Madden, Justin Marks, Michael Pacanowsky, Michael Robillard, Richard Sheridan, Ahmed Sidky, Jon Smart, Stefan Truthän, and Joakim Sundén.

I am grateful for the continuing inspiration offered by the Stoos gathering in 2012, including the insights of Jurgen Appelo, Sanjiv Augustine, Julian Birkinshaw, Rod Collins, Jay Cross, Esther Derby, Peter Hundermark, Klaus Leopold, Uli Loth, Catherine Louis, Melina McKim, Roy Osherove, Deborah Hartmann Preuss, Franz Röösli, Simon Roberts, Michael Spayd, Peter Stevens, John Styffe, Kati Vilkki, and Jonas Vonlanthen.

I am deeply grateful for the conversations with leaders of the Agile movement, including Alex Adamopoulos, Ray Arell, Mike Beedle, Phil

Brock, Mike Cohn, Henrik Esser, Stephen Forte, Peter Green, Bob Hartman, Arjay Hinek, Ron Jeffries, Craig Larman, Evan Leybourn, Heidi Musser, Raj Mudhar, Jeff Sutherland, Ken Schwaber, Bas Vodde, Tom Wessel, and Adrian Zwingli.

I am grateful to Scrum Alliance for the opportunity to serve on its board and the support that it offered for the Learning Consortium in 2015 and the webinar series in 2015–2016.

Much inspiration was gained from Richard and Ilse Straub, and the people I met at the Drucker Forum, including Julian Birkinshaw, Clayton Christensen, Gary Hamel, John Hagel, Bill Fischer, Lynn Forester de Rothschild, Fredmud Malik, Rita Gunther McGrath, Andrew Hill, and Adi Ignatius.

The Innovation for Jobs network has been a tremendous help, including Vint Cerf, David Nordfors, Curt Carlson, and Bob Cohen.

Colleagues at the *Strategy & Leadership* journal, under the courageous leadership of Robert Randall, have been a strong inspiration, including Brian Leavy, Liam Fahey, and Robert Allio.

The global community of management continues to inspire me: Scott Anthony, Joseph Bower, Mike Brittain, Mihir Desai, Bill Lazonick, Harry Moser, Paul Nunes, Lynn Paine, Roger Martin, Stan McChrystal, Carlota Perez, Fred Reichheld, Haydn Shaughnessy, and Anand Venkataraman.

I was fortunate to have many wonderful suggestions from several online review groups. I particularly appreciated the contributions of Matt Anderson, Joel Bancroft-Connors, Alan Barstow, Madelyn Blair, Kas Burger, Felipe Castro, Lisa Cooney, Jeremy Cox, Ed Curley, Stephane Dangel, Charles Dhewa, Gary Douglass, Lyn Dowling, Diane Dromgold, Tony Elmore, Mei Lin Fung, Jay Goldstein, Shane Hastie, Karen Hochberg, Arnaldo Romanos-Hofer, Andrew Holm, John Hovell, Dawna Jones, Cazzy Jordan, Thomas Juli, Michael Kende, Erwin van der Koogh, Yeu Wen Mak, Charles Matthews, Sharon McGann, Imelda McLarnon, Richard Miller, Deborah Mills-Scofield, Dan Montgomery, Steven Moore, Heidi Musser, David Nordfors, Pollyanna Pixton, Peter Randall, Stephen Ritchie, Johanna de Ruyter, Grant Sayer, Nancy Van Schooenderwoert, Robby Slaughter, Erik Smakman, Bruce Smith, Jim

Starrett, Peter Stevens, Rini van Solingen, Carolyn Smithson, Willy Sussland, Nerio Vakil, Tathagat Varma, Germain Verbeemen, Cornelis Vonk, John Warren, Victoria Ward, Robert Ware, Douglas Weidner, Jouw Wijnsma, and Mike Wittenstein.

My agent, Sandra Bond, has been an invaluable sounding board as the book took shape.

Enduring support from my editor at *Forbes*, Fred Allen, has been particularly helpful.

I also recognize the support from my editor at AMACOM, Tim Burgard, whose courage and vision has helped this book to happen.

Amid all this help, I would also like to recognize the most pointed and useful management lessons and insights that have come from my daughter Stephanie and my sister Lyn Dowling.

ABOUT STEVE DENNING

Steve Denning is the author of six business books on leadership, innovation, leadership storytelling, and Agile management, as well as a novel and a volume of poems.

Since 2011, Steve has been writing a leadership column for *Forbes. com*, where he has published more than 700 articles on innovation, Agile management, and the constraints to its implementation. (You can find his articles at blogs.forbes.com/stevedenning.) Steve has been named by A. T. Kearney as "one of the top ten digital innovation influencers on Twitter."

In November 2000, Steve was named by Teleos as one of the world's "ten most admired knowledge leaders." In April 2003, Steve was ranked as one of the world's top 200 business gurus by Thomas Davenport and Laurence Prusak in *What's the Big Idea?* (Harvard Business School Press, 2003).

Steve is a member of the Advisory Board of the Drucker Forum, headquartered in Vienna, Austria, and led sessions at the Drucker Forum in 2014, 2015, and 2016. From 2013 to 2016, he was a member of the board of Scrum Alliance, a nonprofit association of more than 500,000 software developers.

In 2016 and 2017, Steve was named by the Agile Alliance as an "Agile Stalwart." He leads the SD Learning Consortium in which firms, including Barclays, CH Robinson, Cerner, Ericsson, Fidelity Investments, Microsoft, and Vistaprint, are sharing insights on the ongoing transition in making their organizations Agile.

Since 2000, Steve has worked as a consultant and speaker for scores of Fortune 500 companies and government agencies. From 1969 to 2000, he held many management positions at all levels at the World Bank:

- As director of Knowledge Management from 1996 to 2000, he spearheaded a successful strategic shift in the World Bank.

- From 1990 to 1994, he was director of the Southern Africa Department of the World Bank.
- From 1994 to 1996, he was director of the Africa Region of the World Bank.

Business books by Steve Denning include:

- *The Leader's Guide to Radical Management: Reinventing the Workplace for the 21st Century* (Jossey-Bass, 2010). Selected by 800-CEO-READ as one of the best five books on management in 2010.
- *The Secret Language of Leadership: How Leaders Inspire Action Through Narrative* (Jossey-Bass, 2007). Selected by the *Financial Times* as one of the best books of 2007 and by 800-CEO-READ as the best book on leadership in 2007.
- *The Leader's Guide to Storytelling: Mastering the Art and Discipline of Business Narrative* (Jossey-Bass, 1st ed., 2005; 2nd ed., 2011).
- *Squirrel Inc.: A Fable of Leadership Through Storytelling* (Jossey-Bass, 2004).
- *The Springboard: How Storytelling Ignites Action in Knowledge-Era Organizations* (Butterworth Heinemann, 2000).
- Coauthor of *Storytelling in Organizations: How Narrative and Storytelling Are Transforming Twenty-first Century Management* (Elsevier, 2004).

Steve has published more than thirty articles for the management journal *Strategy & Leadership*; the following pieces were named Outstanding Article of the Year:

- "The Reinvention of Management: Part 1: Principles," *Strategy & Leadership* 39, no. 2 (2011).
- "Metrics for the Emerging Creative Economy," *Strategy & Leadership* 42, no. 5 (2014).
- "Understanding the Three Laws of Agile," *Strategy & Leadership* 44, no. 6 (2016, shared prize).

Steve has written four articles for *Harvard Business Review*:

- "Telling Tales," May 2004, https://hbr.org/2004/05/telling-tales.

- "The Internet Is Finally Forcing Management to Care About People," May 2015, https://hbr.org/2015/05/the-internet-is-finally-forcing-management-to-care-about-people.
- "Capitalism's Future Is Already Here," September 2014, https://hbr.org/2014/09/capitalisms-future-is-already-here/.
- "Making Management as Simple as Frisbee," June 2013, https://hbr.org/2013/06/making-management-as-simple-as.

NOTES

INTRODUCTION

1. I am indebted to the following formulation by Alan Murray, who wrote: "It is our belief that the world is in the midst of a new industrial revolution, driven by technology that is connecting everyone and everything, everywhere and all the time, in a vast and intelligent network of interactive data that is creating an economic dynamic increasingly characterized by low or zero marginal costs, massive returns to scale and platform economics." A. Murray, "Six Fundamental Truths About the 21st Century Corporation," *Fortune*, October 22, 2015, http://fortune.com/2015/10/22/six-truths-21st-century-corporation/. My book selects elements of this formulation but puts less emphasis on the competitive advantage created by technology or data and instead gives more emphasis to the different management being used to deploy the technology and the data. It also stresses the outcome for customers and end users: instant, intimate, frictionless value on a large scale. Today, for the most part, technology and data are commodities. See also G. Colvin, "Why Every Aspect of Your Business Is About to Change," *Fortune*, October 22, 2015, http://fortune.com/2015/10/22/the-21st-century-corporation-new-business-models/.

2. Amazon was criticized by analysts for years for missing profit targets, and worse, not even focusing on profits. Now it's bigger than all the other publicly traded retailers put together. Jeff Desjardins, "The Extraordinary Size of Amazon in One Chart," *Business Insider*, January 3, 2017, http://www.businessinsider.com/the-extraordinary-size-of-amazon-in-one-chart-2017-1.

3. Google was founded in 1996 by Larry Page and Sergey Brin. Google's parent company is now Alphabet Inc., an American multinational conglomerate created on October 2, 2015, by the two founders of Google.

4. A. Murray, "The End of Management," *Wall Street Journal*, August 21, 2010, http://www.wsj.com/articles/SB100014240527487044761045754 39723695579664.

5. One key factor: Internally driven innovation and new technology often generate changes that customers don't want or are not willing to pay for.

6. One sign of the scale of the change: Scrum Alliance Inc. has more than a half million members and continues to grow rapidly. The gains of Agile management are not inevitable. It is possible, as later chapters in this book explain, that firms may decide to implement only some, or even none, of the components of Agile management.

7. Many of the elements of Agile management were around long before the Agile Manifesto. Since time immemorial, artists have worked in an iterative fashion: Masterpieces usually evolve through trial and error, rather than emerging perfect from an initial plan. In the nineteenth century, Helmuth von Moltke, the chief of the Prussian (later German) General Staff, developed and applied the concept of Auftragstaktik to cope with uncertainty. Iterative work practices were promoted in the 1930s by Walter Shewhart, a quality expert at Bell Labs. Agile has considerable overlap with design thinking that stems from Herbert A. Simon's book *The Sciences of the Artificial* (Cambridge, MA: MIT Press, 1969). Self-organizing teams have been the staple of new product development for decades. Reducing inventory and delivering value to clients with each iteration are at the heart of lean manufacturing, which was invented by Toyota some fifty years ago. Continuous self-improvement has been a legacy from the total quality movement for more than half a century. Finding ways to measure client delight and the consequent impact on firm growth has been systematically studied by Fred Reichheld and his colleagues at the consulting firm Bain & Company for over twenty-five years.

8. Manifesto for Agile Software Development (http://www.agilemanifesto. org/) is a set of principles for software development in which requirements and solutions evolve through collaboration between self-organizing, cross-functional teams. The full text is included in Box 1-1 at the end of Chapter 1. For some thoughts on the meaning of the values of the Agile Manifesto, see Peter Stevens, "How 'Agile' Are You?," August 31, 2016, https://saat-network.ch/wordpress/wp-content/uploads/2016/08/ Peters-5-Question-Agile-Assessment-RC2.pdf; see also "Five Simple Questions to Determine If You Have the Agile Mindset," *Scrum Breakfast* (blog), August 25, 2016, http://www.scrum-breakfast.com/2016/08/ five-simple-questions-to-determine-if.html.

9. The struggle to reintroduce Agile management to manufacturing is ironic in the sense that many of the historical antecedents of Agile were in manufacturing. See, for example, H. Takeuchi and I. Nonaka, "The New New Product Development Game," *Harvard Business Review*, January 1986, https://hbr.org/1986/01/the-new-new-product-development-game. See also S. Denning, "Transformational Leadership in Agile Manufacturing," *Forbes.com*, August 1, 2012, http://www .forbes.com/sites/stevedenning/2012/08/01/transformational-

leadership-in-agile-manufacturing-wikispeed/. In their book *Scrum* (New York: Crown Publishing, 2014, 35–36), Jeff and J. J. Sutherland describe an illuminating game that teaches the role of Agile management in manufacturing through building paper airplanes by way of W. Edwards Deming's cycle of Plan, Do, Check, Act (PDCA).

10. S. Denning, "The Best-Kept Management Secret on the Planet," *Forbes.com*, April 12, 2012, http://www.forbes.com/sites/stevedenning/2012/04/09/the-best-kept-management-secret-on-the-planet-agile/.

11. Innosight, "Creative Destruction Whips Through Corporate America: S&P 500 Lifespans Are Shrinking," February 2012, http://www.innosight.com/innovation-resources/strategy-innovation/upload/creative-destruction-whips-through-corporate-america_final2015.pdf.

12. Martin Reeves, Simon Levin, Daichi Ueda, "The Biology of Corporate Survival," *Harvard Business Review*, January–February 2016, https://hbr.org/2016/01/the-biology-of-corporate-survival.

13. M. Andreesen, "Why Software Is Eating the World," *Wall Street Journal*, August 20, 2011, https://www.wsj.com/articles/SB10001424053111903480904576512250915629460.

14. The SD Learning Consortium (SDLC) is a nonprofit corporation registered in Virginia. The author is an unpaid pro bono director of the SDLC. The 2016 report of the SDLC titled "The Entrepreneurial Organization at Scale" is available at http://sdlearningconsortium.org/index.php/home/what-we-have-learned/full-report-2016/. A predecessor of the SD Learning Consortium was sponsored by Scrum Alliance in 2015.

15. On November 18, 2016, Julian Birkinshaw, professor of Strategy and Entrepreneurship at the London Business School and director of the Deloitte Institute of Innovation and Entrepreneurship, declared provocatively that we are living in "the Age of Agile." S. Denning, "The Age of Agile: What Every CEO Needs to Know," *Forbes.com*, December 9, 2016, https://www.forbes.com/sites/stevedenning/2016/12/09/the-age-of-agile-what-every-ceo-needs-to-know/.

CHAPTER 1

1. P. Drucker, *Innovation and Entrepreneurship* (New York: Harper & Row, 1985), 313.

2. C. Johnson, "Who Are We?? Chris Johnson from Idea to Execution: Spotify's Discover Weekly," *SlideShare*, November 15, 2015, http://www.slideshare.net/MrChrisJohnson/from-idea-to-execution-spotifys-discover-weekly/2-Who_are_WeChris_Johnson_Edward.

3. Adam Pasick, "The Magic That Makes Spotify's Discover Weekly

Playlists So Damn Good," *Quartz*, December 21, 2015, http://qz.com/571007/the-magic-that-makes-spotifys-discover-weekly-playlists-so-damn-good/.

4. Ibid.

5. Michael Harte, "Digital Transformation in Banking," YouTube video, 19:49, July 13, 2015, https://www.youtube.com/watch?v=d6mqxcevZj0.

6. D. K. Rigby, J. Sutherland, and H. Takeuchi, "Embracing Agile," *Harvard Business Review*, April 2016, https://hbr.org/2016/05/embracing-agile.

7. M. Lurie, "The Five Disciplines of Agile Organizations," in *Agility Hackathon E-book: Compilation of Participants' Experiences and Learnings*, McKinsey & Company, April 2016.

8. The report of the 2016 site visits by the SD Learning Consortium is available at http://sdlearningconsortium.org/index.php/home/what-we-have-learned/full-report-2016/. The report of the 2015 site visits of the Learning Consortium organized by the Scrum Alliance is available at http://sdlearningconsortium.org/index.php/home/what-we-have-learned/full-report-2015/.

9. A. Murray, "Six Fundamental Truths About the 21st Century Corporation," *Fortune*, October 22, 2015, http://fortune.com/2015/10/22/six-truths-21st-century-corporation/.

10. The Ericsson example was first published in "The Entrepreneurial Organization at Scale," Report of the SD Learning Consortium, November 9, 2016, 3, http://sdlearningconsortium.org/wp-content/uploads/Report-r28-NOV-9-2016-PUBLIC-VERSION.pdf, and is reproduced here under the Creative Commons license.

11. The forty Agile methods delineated by Craig Smith are depicted in a clever graphic by Australian designer Lynne Cazaly, in Steve Denning, "Explaining Agile," *Forbes.com*, September 8, 2016, http://www.forbes.com/sites/stevedenning/2016/09/08/explaining-agile/.

12. General Stanley McChrystal, *My Share of the Task: A Memoir* (New York: Penguin Publishing Group, Kindle Edition, 2013); Mihály Csíkszentmihályi, *Creativity: Flow and the Psychology of Discovery and Invention* (New York: Harper Perennial, 1996).

13. J. Clifton, "Workplace Disruption: From Annual Reviews to Coaching," *Gallup.com*, February 15, 2017, http://www.gallup.com/opinion/chairman/203876/workplace-disruption-annual-reviews-coaching.aspx.

14. E. Schmidt and J. Rosenberg, *How Google Works* (New York: Grand Central Publishing, 2014), 86.

15. J. Kotter, *Accelerate* (Boston: Harvard Business Review Press, 2014).

16. For dual operating systems, see Kotter, Accelerate, and S. D. Anthony, C. G. Gilbert, and M. W. Johnson, Dual Transformation: How to

Reposition Today's Business While Creating the Future (Boston: Harvard Business Review Press, 2017). For knowledge funnels with design thinking, see R. Martin, The Design of Business (Boston: Harvard Business Review Press, 2009). Martin describes a knowledge funnel by which organizations progressively figure out mysteries, which then turn into business heuristics that can be exploited with the addition of judgment, and eventually algorithms, which can be exploited precisely.

17. Alvin Toffler, *Future Shock* (New York: Random House, 1970), 10–11.

CHAPTER 2

1. E. F. Schumacher, *Small Is Beautiful: Economics as if People Mattered* (London: Blond & Briggs, 1973), 259.

2. H. Shaughnessy, *The Elastic Enterprise: The New Manifesto for Business Revolution* (Dublin, Ohio, Telemachus Press, 2012).

3. "The Newton" refers to a series of personal digital assistants (PDAs) developed and marketed by Apple Inc. Apple launched the platform in 1987 and shipped the first devices in 1993. Production officially ended on February 27, 1998. According to former Apple CEO John Sculley, the corporation invested approximately US$100 million to develop the Newton.

4. R. Stross, "Billion Dollar Flop: Airforce Stumbles on Software Plan," *New York Times*, December 9, 2012, http://www.nytimes.com/ 2012/12/09/technology/air-force-stumbles-over-software-modernization- project.html. See also S. Denning, "Reconciling Innovation with Control: A $1.3 Billion Lesson in Agile," *Forbes.com*, December 11, 2012, https://www.forbes.com/sites/stevedenning/2012/12/11/reconciling- innovation-with-control-the-air-forces-1-3-billion-lesson-in-agile/.

5. Stross, "Billion Dollar Flop."

6. Stross, Ibid.

7. "SAAB JAS-39 Sweden (Super Fighter)," YouTube video, 2:51, October 13, 2011, https://www.youtube.com/watch?v=SOw0Og0i8pA.

8. S. Joshi, "Gripen Operational Cost Lowest of All Western Fighters: Jane's," *StratPost*, July 4, 2012, http://www.stratpost.com/ gripen-operational-cost-lowest-of-all-western-fighters-janes.

9. Ibid.

10. B. Sweetman, " Is Saab's New Gripen the Future of Fighters?," *Aviation Week and Space Technology*, March 24, 2014, http://aviation- week.com/defense/saab-s-new-gripen-future-fighters.

11. J. Hirsch, "Elon Musk: Model S Not a Car but a 'Sophisticated Computer on Wheels," *LA Times*, March 19, 2015, http://www.latimes .com/business/autos/la-fi-hy-musk-computer-on-wheels-20150319- story.html.

12. S. Denning, *The Leader's Guide to Radical Management* (San Francisco: Jossey-Bass, 2010), 118–121.

13. H. Takeuchi and I. Nonaka, "The New New Product Development Game," *Harvard Business Review*, January 1986, https://hbr.org/1986/01/the-new-new-product-development-game.

14. J. P. Womack and D. T. Jones, *The Machine That Changed the World: The Story of Lean Production—Toyota's Secret Weapon in the Global Car Wars That Is Now Revolutionizing World Industry* (New York: Free Press, 1990). The book recorded extraordinary differences in outcomes between factories using Toyota's approach (which the book called "lean") and those run on traditional lines (which it called "mass manufacturing"). On the design side, the lean approach was more productive on every measurable aspect. It wasn't a difference between Japanese factories and U.S. factories. In fact, some of the best factories were in the United States and some of the worst were in Japan. What made the difference was how the factory was run. Nor was it a question of who was running the plant. In the study, the top-rated plant in terms of quality and productivity wasn't a Japanese plant at all. It was a Ford plant in Hermosillo, Mexico.

15. These site visits took place in 2015 under the auspices of the Learning Consortium for the Creative Economy organized by Scrum Alliance, and in 2016 and in 2017 under the auspices of the SD Learning Consortium.

16. J. Rozovsky, "The Five Keys to a Successful Google Team," *re:Work* (blog), November 17, 2015, https://rework.withgoogle.com/blog/five-keys-to-a-successful-google-team/; C. Duhigg, "What Google Learned from Its Quest to Build the Perfect Team," *New York Times*, February 25, 2016, https://www.nytimes.com/2016/02/28/magazine/what-google-learned-from-its-quest-to-build-the-perfect-team.html.

17. S. Denning, "The Joy of Work: Menlo Innovations," Forbes.com, August 2, 2016, http://www.forbes.com/sites/stevedenning/2016/08/02/the-joy-of-work-menlo-innovations/.

18. D. H. Pink, *Drive* (New York: Riverhead Books, 2009).

19. S. Denning, "From CEO Takers to CEO Makers," *Forbes.com*, August 20, 2014, http://www.forbes.com/sites/stevedenning/2014/08/20/from-ceo-takers-to-ceo-makers-the-great-transformation/.

20. In general, the Agile movement has avoided "New Age" talk about "the next stage of human consciousness," as promoted by Frederic Laloux's *Reinventing Organizations: A Guide to Creating Organizations Inspired by the Next Stage of Human Consciousness* (Nelson Parker, 2014). It also stayed away from the linguistic extremes of the holacracy movement that sometimes talks of doing away with job titles,

managers, and hierarchy. See S. Denning, "No Managers, No Hierarchy, No Way," *Forbes.com*, April 18, 2014, https://www.forbes.com/sites/stevedenning/2014/04/18/no-managers-no-hierarchy-no-way/; and S. Denning, "Making Sense of Zappos and Holacracy," *Forbes.com*, January 15, 2015, https://www.forbes.com/sites/stevedenning/2014/01/15/making-sense-of-zappos-and-holacracy/; S. Denning, "Is Holacracy Succeeding at Zappos," *Forbes.com*, May 23, 2015, https://www.forbes.com/sites/stevedenning/2015/05/23/is-holacracy-succeeding-at-zappos/. Hierarchy still exists in Agile management, but it is for the most part a hierarchy of competence, not a hierarchy of authority.

CHAPTER 3

1. "Attribution of Schopenhauer's Three Stages of Truth," discussion in "Quotes Debunked," November 5, 2012, https://www.metabunk.org/attribution-of-schopenhauers-three-stages-of-truth.t897/.

2. Copernicus was not the first person to formulate the sun-centered view of the universe. Aristarchus of Samos (c. 310–c. 230 BC), an ancient Greek astronomer and mathematician, presented the first known heliocentric model. Claudius Ptolemy preferred the geocentric model that dominated until the heliocentric theory was successfully revived by Copernicus, after which Johannes Kepler described planetary motions with greater accuracy with what are known as Kepler's laws, and Isaac Newton gave a theoretical explanation based on laws of gravitational attraction and dynamics.

3. The plausibility of Roman Catholic theology and the Divine Right of Kings was implicitly dependent in part on the notion that the earth is the physical center of the universe, thereby warranting the attentions of a Divine Being. As it became apparent that the earth is a tiny speck of dust in an infinitely vast universe, that plausibility came under more intensive questioning.

4. T. Kuhn, *The Copernican Revolution: Planetary Astronomy in the Development of Western Thought* (Boston: Harvard University Press, 1957), 94.

5. P. Drucker, *The Practice of Management* (New York: HarperCollins, 1954), 37.

6. Ibid.

7. M. Jensen and K. Murphy, "CEO Incentives—It's Not How Much You Pay, But How," *Harvard Business Review*, May 1990, https://hbr.org/1990/05/ceo-incentives-its-not-how-much-you-pay-but-how.

8. R. Gulati, *Reorganize for Resilience* (Boston: Harvard Business School Press, 2010).

9. K. R. Jamison, *Exuberance: The Passion for Life* (New York: Knopf, 2004), 5.

10. J. Clifton, "Workplace Disruption: From Annual Reviews to Coaching," *Gallup.com*, February 15, 2017, http://www.gallup.com/opinion/chairman/203876/workplace-disruption-annual-reviews-coaching.aspx.

11. A. Koller, "Stephen Fry on Things He Had Learned in Life," *Design Research* (blog), January 17, 2013, http://blog.andreaskoller.com/2013/01/stephen-fry-on-things-he-has-learned-in-life/.

12. S. Denning, *The Leader's Guide to Radical Management* (San Francisco: Jossey-Bass, 2010), 118-120.

13. T. L. Friedman, *That Used to Be Us: How America Fell Behind in the World It Invented and How We Can Come Back* (New York: Farrar, Straus and Giroux, 2011), 95.

14. "Scott Galloway at DLC 2017," YouTube video, 23:06, January 20, 2017, https://www.youtube.com/watch?v=cFxdgZ1az9s&feature=youtu.be&t=15s.

15. S. Galloway, "Alexa: How Can We Kill Brands?," *No Mercy/No Malice*, May 12, 2017, https://www.l2inc.com/no-mercy-no-malice/alexa-how-can-we-kill-brands.

16. In traditional top-down bureaucracies, with multiple vertical layers of authority and many different departments and divisions, work jams are occurring all over the organization on a daily basis, though typically, no one recognizes them or does anything about them. Work sits waiting in queues. Approvals hold things up. Customers try to get answers and wait for responses. Well-intended cost savings implemented in one part of the organization are slowing things down in another part of the organization, retarding the overall delivery of value to customers. Big production runs are particularly problematic, because they maximize work in process and inventory, generating direct costs of working capital and warehousing, hiding quality problems, and causing noxious secondary effects.

17. G. Stalk, "Time—The Next Source of Competitive Advantage," *Harvard Business Review*, July 1988, https://hbr.org/1988/07/time-the-next-source-of-competitive-advantage.

18. P. Noonan, "A Caveman Won't Beat a Salesman," *Wall Street Journal*, November 18, 2011, https://www.wsj.com/articles/SB10001424052970203611404577044613194688678.

19. J. Tapper, "General Electric Paid No Federal Taxes in 2010," *ABC-News.com*, March 25, 2011, http://abcnews.go.com/Politics/general-electric-paid-federal-taxes-2010/story?id=13224558.

20. S. Denning, "Retirement Heist: How Firms Plunder Workers' Nest Eggs," *Forbes.com*, October 19, 2011, https://www.forbes.com/sites/

stevedenning/2011/10/19/retirement-heist-how-firms-plunder-workers-nest-eggs/.

21. S. Denning, "Why Are Fannie and Freddie CEOs Paid So Much?," *Forbes.com*, November 16, 2011, https://www.forbes.com/sites/stevedenning/2011/11/16/why-are-fannie-freddie-ceos-paid-so-much/.

22. S. Denning, "Why Amazon Can't Make a Kindle in the U.S.A.," *Forbes.com*, August 17, 2011, https://www.forbes.com/sites/stevedenning/011/08/17/why-amazon-cant-make-a-kindle-in-the-usa/.

23. S. Denning, "Resisting the Lure of Short-Termism," *Forbes.com*, January 8, 2017, https://www.forbes.com/sites/stevedenning/2017/01/08/resisting-the-lure-of-short-termism-how-to-achieve-long-term-growth/.

24. Denning, "Why Are Fannie and Freddie CEOs Paid So Much?" By creating the basis for questioning the legitimacy of compensation for corporate leaders, the Law of the Customer enables the true financial and social worth of corporate leaders to be examined and recalibrated. Top managers will continue to exist, but the extraordinary compensation and prerogatives of the C-suite will come under increasing scrutiny, both in absolute terms of what the leaders themselves are contributing and also relative to creative talent that is truly adding real value to customers.

25. T. Kuhn, *The Structure of Scientific Revolutions* (Chicago: University of Chicago Press, 1962).

26. F. W. Taylor, The *Principles of Scientific Management* (New York: Harper & Brothers, 1911), 7.

CHAPTER 4

1. S. McChrystal, T. Collins, D. Silverman, and C. Fussell, *Team of Teams: New Rules of Engagement for a Complex World* (New York: Penguin Publishing Group, 2015), 71–72.

2. McChrystal et al., *Team of Teams*, 69–70.

3. Ibid., 127.

4. Ibid., 118.

5. Ibid., 122.

6. Ibid., 123.

7. Ibid., 127.

8. Ibid., 127–128.

9. Ibid., 156.

10. Ibid., 84.

11. S. Denning, "The Key Missing Leadership Ingredient: Part 2: The Military," *Forbes.com*, July 31, 2012, https://www.forbes.com/sites/stevedenning/2012/07/31/the-key-missing-leadership-ingredient-part-2-the-military/. "Mission command" over "detailed command" has

been championed by the U.S. military, but overall, the military structure remains top-down and bureaucratic. See Chapter 11.

12. McChrystal et al., *Team of Teams*, 159.

13. Ibid., 161.

14. Ibid., 175–176.

15. Ibid.

16. Ibid., 225–256.

17. Ibid., 243. In management matters, there is always a risk of apparent oversimplification, as complex events risk being attributed to a single cause. There are always multiple causes and multiple outcomes. The ultimate results are the convergence of many separate and independent developments and can lead to diverse effects that can go for generations.

18. The U.S. Task Force's network in Iraq helped to produce tactical "wins" in terms of the capture of former president Saddam Hussein and the killing of the extremist leader Abu Musab al Zarqawi. However, in broader terms, the story of General McChrystal's Task Force in Iraq does not show that the Law of the Network by itself is any kind of panacea in a complex political situation. The Task Force's network was a better organizational arrangement for achieving the goal that had been assigned to it by the U.S. government, but was the goal itself sound? The U.S. government had decided to invade Iraq, in pursuit of "weapons of mass destruction," which turned out not to exist. In May 2003, shortly after the initial invasion, United States president George W. Bush, in front of a sign that declared "Mission Accomplished," declared an "end to major combat operations in Iraq." As guerilla warfare in Iraq steadily took shape, commanders on the ground knew that they were dealing with an insurgency; see S. McChrystal, *My Share of the Task* (New York: Penguin, 2013), 122. Yet at higher levels, the U.S. government was reluctant to accept that fact. Secretary of Defense Donald Rumsfeld even refused for several years to allow the use of the term "insurgency." The U.S. government's overall strategy remained one of finding and killing "bad guys." The Task Force in Iraq was successfully doing that. It was killing more, capturing more, and its facilities were full of thousands of detainees. But no matter how many "bad guys" the Task Force killed, there were more to take their place. In this regard, the accidental killing of innocent civilians, the harsh interrogations of Iraqi detainees, and the horrifying abuses that occurred at the Abu Ghraib prison obviously didn't help. The U.S. government strategy at the time was paying insufficient attention to the Law of the Customer: Ultimate success would depend on winning the hearts or minds of the Iraqi people. "The classic doctrine, which was

developed by the British in Malaya in the nineteen-forties and fifties, says that counterinsurgency warfare is twenty per cent military and eighty per cent political," writes analyst George Packer. "The focus of operations is on the civilian population: isolating residents from insurgents, providing security, building a police force, and allowing political and economic development to take place so that the government commands the allegiance of its citizens. A counterinsurgency strategy involves both offensive and defensive operations, but there is an emphasis on using the minimum amount of force necessary. For all these reasons, such a strategy is extremely hard to carry out, especially for the American military, which [focuses] on combat operations. Counterinsurgency cuts deeply against the Army's institutional instincts." The U.S. Task Force in Iraq had been assigned the "20 percent military" part of the mission, but the U.S. government never fully embraced, or committed the resources necessary for, accomplishing the "80 percent political" part of the mission (George Packer, "The Lesson of Tal Afar," *The New Yorker*, April 10, 2006, http://www.newyorker.com/magazine/2006/04/10/the-lesson-of-tal-afar). Yet despite lack of support from other parts of the government, operations in Iraq were demonstrating the potential of a "counterinsurgency" strategy: General James Mattis in Fallujah, Colonel H. R. McMaster in Tal Afar near the Syrian border, and General David Petraeus in Mosul began putting a focus on securing the population and bringing economic development, not just killing "the enemy." They sent troops into the cities and kept them there, establishing connections with local leaders and Iraqi Army units, gathering intelligence from the local population on the extremists, providing security in the streets, and working with the warring factions. Progress was significant, although the gains didn't last after these commanders moved on, as the U.S. government was slow to embrace and sustain full support for a counterinsurgency strategy. See F. Kaplan, *The Insurgents: David Petraeus and the Plot to Change the American Way of War* (New York: Simon & Schuster, 2013), 303; and McChrystal, *My Share of the Task*, 129. Organizational success ultimately depends on both "doing the right thing" and "doing it right." In mastering the Law of the Network, the U.S. Army Task Force got closer to "doing it right" in terms of accomplishing its part of the mission. But that was not enough for broader political success when the overall U.S. government strategy was flawed.

19. Carlota Perez, *Technological Revolutions and Financial Capital: The Dynamics of Bubbles and Golden Ages* (Cheltenham, U.K.: Edward Elgar, 2003), 21. The network arrangement is, Perez writes, "replicated in many other organizations confronted with a large and complex

task in government, in hospitals, in universities, in trade unions and political parties. In the West and in the Soviet system, in developed and developing countries."

20. Alcoholics Anonymous, General Service Office, "A.A. Fact File," June 2013, http://www.aa.org/assets/en_US/m-24_aafactfile.pdf.

21. M. Gladwell, "The Cellular Church," *The New Yorker*, September 12, 2005, http://www.newyorker.com/magazine/2005/09/12/the-cellular-church.

22. S. Denning, "John Hagel's Wake-up Call for Business," *Forbes.com*, January 2, 2017, https://www.forbes.com/sites/stevedenning/2017/01/02/john-hagels-wake-up-call-for-business-how-to-launch-a-change-movement/.

23. A top-down, all-or-nothing approach may create a backlash, so even if a company appears to succeed, the sustainability of the effort is likely to be in question. The cost of victory may be too high. There's also a risk that a poorly implemented change will discredit the very idea of managing without bureaucracy for a long time to come.

24. S. Denning, "Gary Hamel's $3 Trillion Prize for Killing Bureaucracy," *Forbes.com*, March 29, 2016, https://www.forbes.com/sites/stevedenning/2016/03/29/gary-hamels-3-trillion-prize-for-killing-bureaucracy/.

CHAPTER 5

1. B. Harry, "Agile Project Management in Visual Studio ALM V.Next," *Brian Harry's Blog*, June 14, 2011, https://blogs.msdn.microsoft.com/bharry/2011/06/14/agile-project-management-in-visual-studio-alm-v-next/.

2. K. Schwaber, "Microsoft and Brian Harry," *Ken Schwaber's Blog: Telling It Like It Is*, July 18, 2011, https://kenschwaber.wordpress.com/2011/07/18/microsoft-and-brian-harry/.

3. B. Harry, "Self-forming Teams at Scale," *Brian Harry's Blog*, July 24, 2015, http://blogs.msdn.com/b/bharry/archive/2015/07/24/self-forming-teams-at-scale.aspx.

CHAPTER 6

1. L. Freedman, *Strategy: A History* (Oxford, U.K.: Oxford University Press, 2013), ix.

2. S. Anthony, "What Do You Really Mean by Business 'Transformation,'" *Harvard Business Review*, February 29, 2016, https://hbr.org/2016/02/what-do-you-really-mean-by-business-transformation.

3. C. M. Christensen, K. Dillon, T. Hall, and D. Duncan, *Competing Against Luck: The Story of Innovation and Customer Choice* (New York: HarperCollins, 2016), 37.

4. D. Brooks, "The Creative Monopoly," *New York Times*, April 24,

2012, http://www.nytimes.com/2012/04/24/opinion/brooks-the-creative-monopoly.html.

5. P. Vlaskovits, "Henry Ford, Innovation, and That 'Faster Horse' Quote," *Harvard Business Review*, August 2011, https://hbr.org/2011/08/henry-ford-never-said-the-fast. There is no evidence that Henry Ford ever made this statement.

6. W. C. Kim and R. Mauborgne, *Blue Ocean Strategy, Expanded Edition: How to Create Uncontested Market Space and Make the Competition Irrelevant* (Boston: Harvard Business Review Press, 2015).

7. A/B testing involves comparing two versions of a web page to find out which one performs better. The two variants are shown to different visitors at the same time. The test enables developers to determine which version gives the better response rate.

8. Kim and Mauborgne, *Blue Ocean Strategy*, 40.

9. There is now even a movement afoot called "anti-lean-startups," which makes the case for adopting a different approach for major innovation that involves scientific breakthroughs; see D. Mortensen, "Why Anti-Lean Startups Are Back," *LinkedIn Pulse*, November 2, 2016, https://www.linkedin.com/pulse/why-anti-lean-startups-back-dennis-r-mortensen.

10. N. Schwieters, "The End of Conventional Industry Sectors," *strategy + business*, January 3, 2017, https://www.strategy-business.com/blog/The-End-of-Conventional-Industry-Sectors.

11. Christensen et al., *Competing Against Luck*, 89.

12. C. R. Carlson and W. W. Wilmot, *Innovation: The Five Disciplines for Creating What Customers Want* (New York: Crown Business, 2006).

13. C. Carlson, "'Do You Have a Value-Creation Playbook?' 'No,'" *Drucker Forum Blog*, December 7, 2016, https://www.druckerforum.org/blog/?p=1429.

14. A. Venkataraman, "Can Innovation Be Learned or Taught?," *Quora*, January 20, 2017, https://www.quora.com/Can-innovation-be-learned-or-taught/answer/Anand-Venkataraman.

15. W. C. Kim and R. Mauborgne, *Blue Ocean Strategy, Expanded Edition: How to Create Uncontested Market Space and Make the Competition Irrelevant* (Boston: Harvard Business Review Press, 2015), 90.

16. Venkataraman, "Can Innovation Be Learned or Taught?"

17. Shaughnessy quoted in S. Denning, "How Apple Achieves Massive Scale Without Pain," *Forbes.com*, September 3, 2013, https://www.forbes.com/sites/stevedenning/2013/09/03/how-the-elastic-enterprise-defeats-the-sclerosis-of-scale/.

18. Schwieters, "The End of Conventional Industry Sectors."

19. Shaughnessy quoted in S. Denning, "How the Elastic Enterprise Defeats the Sclerosis of Scale," *Forbes.com*, September 3, 2013, https://www.forbes.com/sites/stevedenning/2013/09/03/how-the-elastic-

enterprise-defeats-the-sclerosis-of-scale/.

20. S. Denning, "John Hagel's Wake-up Call for Business," *Forbes.com*, January 2, 2017, https://www.forbes.com/sites/stevedenning/2017/01/ 02/john-hagels-wake-up-call-for-business-how-to-launch-a-change- movement/.

21. Venkataraman, "Can Innovation Be Learned or Taught?"

22. Ibid.

23. Ibid.

CHAPTER 7

1. L. Gerstner, *Who Says Elephants Can't Dance?* (New York: HarperCollins, 2003), 182.

2. W. Wilmot and J. Hocker, *Interpersonal Conflict* (New York: McGraw-Hill, 2010). Wilmot suggested that he and Carlson write a book together about the innovation model Carlson was introducing to SRI. See C. R. Carlson and W. W. Wilmot, *Innovation: The Five Disciplines for Creating What Customers Want* (New York: Crown Business, 2006).

3. SRI's Innovation-for-Impact Playbook was described in C. Carlson and W. Wilmot, *Innovation: The Five Disciplines for Creating What Customers Want*. See also H. Kressel and N. Winarsky, *If You Really Want to Change the World: A Guide to Creating, Building, and Sustaining Breakthrough Ventures* (Boston: Harvard Business Review Press, 2015).

4. See S. Denning, *The Leader's Guide to Storytelling, 2nd ed.* (San Francisco: Jossey-Bass, 2011); S. Denning, "The Four Stories You Need to Lead Deep Organizational Change," *Forbes.com*, July 25, 2011; http:// www.forbes.com/sites/stevedenning/2011/07/25/the-four-stories-you- need-to-lead-deep-organizational-change/; S. Denning, "How to Say No While Also Inspiring People," *Forbes.com*, May 30, 2011, https://www .forbes.com/sites/stevedenning/2011/05/30/leadership-how-to-say-no- while-also-inspiring-people/.

PART TWO: MANAGEMENT TRAPS

1. "BlackRock CEO Larry Fink tells the world's biggest business leaders to stop worrying about short-term results," *BusinessInsider.com*, April 14, 2015, http://www.businessinsider.com/larry-fink-letter-to-ceos-2015-4. In January 2015, Fink's annual letter to CEOs stated: "Companies have begun to devote greater attention to these issues of long-term sustainability, but despite increased rhetorical commitment, they have continued to engage in buybacks at a furious pace." BlackRock Inc., "Open Letter to S&P 500 CEOs," https://www.blackrock.com/corporate/en-us/ investor-relations/larry-fink-ceo-letter.

2. A historical perspective is sobering. In his book *The Age of Heretics*,

Art Kleiner wrote about the corporate mavericks of the 1950s, 1960s, and 1970s who pioneered self-managing work teams. But most of their work didn't stick. Yet today, the situation is different. The cost of bureaucracy—in terms of missed opportunities, underleveraged talents, and squandered innovation—is much higher. The competitive environment is much tougher. So, the importance of embracing Agile management is much higher than it was in the 1950s to 1970s. What was once an interesting experimental option is now a market-driven necessity.

CHAPTER 8

1. F. Guerrera, "Welch Condemns Share Price Focus," *Financial Times*, March 12, 2009, https://www.ft.com/content/294ff1f2-0f27-11de-ba10-0000779fd2ac#axzz1eiLpL2PZ; S. Denning, "The World's Dumbest Idea: Maximizing Shareholder Value," *Forbes.com*, November 28, 2011, http://www.forbes.com/sites/stevedenning/2011/11/28/maximizing-shareholder-value-the-dumbest-idea-in-the-world/.

2. R. Martin, *Fixing the Game: Bubbles, Crashes, and What Capitalism Can Learn from the NFL* (Boston: Harvard Business Review Press, 2012), 20. Denning, "The World's Dumbest Idea."

3. A rare exception to the total focus in the NFL on playing the game in accordance with the rules was the controversy involving the allegation that the New England Patriots deliberately underinflated footballs used in their victory against the Indianapolis Colts in the American Football Conference Championship Game of the 2014–2015 NFL playoffs. The controversy resulted in Patriots quarterback Tom Brady being suspended for four games and the team being fined $1 million and losing two draft picks.

4. J. R. Graham, C. R. Harvey, and S. Rajgopal, "Value Destruction and Financial Reporting Decisions," *Financial Analysts Journal* 62, no. 6 (2006), http://www.cfapubs.org/doi/10.2469/faj.v62.n6.4351, cited in G. Mukunda, "The Price of Wall Street's Power," *Harvard Business Review*, June 2014, https://hbr.org/2014/06/the-price-of-wall-streets-power.

5. See R. Martin, *Fixing the Game*, 97.

6. Financial Accounting Standards Board (FASB) Regulation 142 (2002).

7. "Analyse This: The Enduring Power of the Biggest Idea in Business," *The Economist*, March 31, 2016, http://www.economist.com/news/business/21695940-enduring-power-biggest-idea-business-analyse.

8. "The Error at the Heart of Corporate Leadership," *Harvard Business Review*, May–June 2017.

9. Guerrera, "Welch Condemns Share Price Focus"; Denning, "The World's Dumbest Idea."

10. S. Denning, "The Hegemony of Shareholder Value Is Finally Ending," *Forbes.com*, August 29, 2012, http://www.forbes.com/sites/stevedenning/

2012/08/29/is-the-hegemony-of-shareholder-value-finally-ending/. The "garbage can" theory of organizations was developed to describe the mid-twentieth-century university by a trio of business school professors—Michael D. Cohen, James G. March, and Johan P. Olsen—but it had wider applicability to many large corporations. As goals were unclear, systems, rules, and procedures that had been devised to achieve financial and administrative order took on a life of their own. The processes themselves became the main preoccupation of those working within the organization. As employees oriented their lives around the processes, the purpose of the organization could get lost in a fog of bureaucracy. See M. D. Cohen, J.G. March, and J.P. Olsen, "A Garbage Can Model Of Organizational Choice," *Administrative Science Quarterly*, March 1972, Vol 17, No. 1, 1-25.

11. See W. Kiechel, *Lords of Strategy* (Boston: Harvard Business Review Press, 2013), 246, on "the Four Horsemen of the Apocalypse."

12. M. Friedman, *Capitalism and Freedom* (Chicago: University of Chicago Press, 1962), 133.

13. M. Friedman, "The Social Responsibility of Business Is to Increase Its Profits," *New York Times Magazine*, September 13, 1970, http://www. colorado.edu/studentgroups/libertarians/issues/friedman-soc-resp-business.html.

14. How did the future Nobel Prize winner arrive at these conclusions? In a paper that accuses others of "analytical looseness and lack of rigor," it's curious that the paper assumes its conclusion at the outset. "In a free-enterprise, private-property system," the article states flatly as an obvious truth requiring no justification or proof, "a corporate executive is an employee of the owners of the business"—namely, the shareholders. As a matter of law, this is incorrect. The executive is an employee of the corporation. In Friedman's imaginary world, the executive and the stockholders have "a voluntary contractual arrangement" that is "clearly defined," even though in the real world, no such contractual arrangement exists: The executive's legal contract is with the corporation. In Friedman's imaginary world of economic models, an organization is a mere "legal fiction." Friedman defines the corporation out of existence in order to prove his predetermined conclusion.

15. M. C. Jensen and W. H. Meckling, "Theory of the Firm: Managerial Behavior, Agency Costs and Ownership Structure," *Journal of Financial Economics* 3. no. 4 (1976), http://papers.ssrn.com/sol3/papers.cfm?abstract_id=94043.

16. M. C. Jensen and K. J. Murphy, "CEO Incentives—It's Not How Much You Pay, But How," *Harvard Business Review*, May–June 1990,

https://hbr.org/1990/05/ceo-incentives-its-not-how-much-you-pay-but-how.

17. S. Gandel, "What Caused Valeant's Epic 90% Plunge?," Reuters, March 20, 2016, http://fortune.com/2016/03/20/valeant-timeline-scandal/.

18. P. Henning, "When Money Gets in the Way of Corporate Ethics," *New York Times*, April 17, 2017, https://www.nytimes.com/2017/04/17/business/dealbook/when-money-gets-in-the-way-of-corporate-ethics.html.

19. Associated Press, "Volkswagen executive pleads guilty in emissions scandal," *Los Angeles Times*, August, 4, 2017, http://www.latimes.com/business/la-fi-hy-volkswagen-emissions-guilty-20170804-story.html.

20. See these articles by S. Denning: "Retirement Heist: How Firms Plunder Workers' Nest Eggs," *Forbes.com,* October 19, 2011, http://www.forbes.com/sites/stevedenning/2011/10/19/retirement-heist-how-firms-plunder-workers-nest-eggs/; "GE Discusses Retirement Heist," *Forbes.com*, October 21, 2011, http://www.forbes.com/sites/stevedenning/2011/10/21/ge-discusses-retirement-heist/Heist Part 3," *Forbes.com*, October 22, 2011, http://www.forbes.com/sites/stevedenning/2011/10/22/retirement-heist-part-3-ellen-schultz-replies-to-ge/; "How Your Pension Got Turned into Scotch or Cheese," *Forbes.com*, April 22, 2013, http://www.forbes.com/sites/stevedenning/2013/04/22/sorry-about-your-pension-scotch-cheese-or-golf/.

21. J. Asker, J. Farre-Mensa, and A. Ljungqvist, "Corporate Investment and Stock Market Listing: A Puzzle?," *Review of Financial Studies* 28, no. 2 (February 2015): 342–390, http://papers.ssrn.com/sol3/papers.cfm?abstract_id=1603484; S. Denning, "How CEOs Became Takers, Not Makers," *Forbes.com*, August 18, 2014, http://www.forbes.com/sites/stevedenning/2014/08/18/hbr-how-ceos-became-takers-not-makers/.

22. S. Denning, "How the World's Dumbest Idea Killed the Economic Recovery," *Forbes.com*, July 29, 2013, http://www.forbes.com/sites/stevedenning/2013/07/29/how-the-worlds-dumbest-idea-killed-the-us-economic-recovery/; S. Denning, "Do We Need a Revolution in Management?," *Forbes.com*, May 26, 2014, http://www.forbes.com/sites/stevedenning/2014/05/26/clayton-christensen-do-we-need-a-revolution-in-management/.

23. J. Wiens and C. Jackson, "The Importance of Young Firms for Economic Growth," The Kauffman Foundation, September 13, 2015, http://www.kauffman.org/what-we-do/resources/entrepreneurship-policy-digest/the-importance-of-young-firms-for-economic-growth; S.

Denning, "The Surprising Truth About Where New Jobs Come From," *Forbes.com*, October 29, 2014, https://www.forbes.com/sites/stevedenning/2014/10/29/the-surprising-truth-about-where-new-jobs-come-from/.

24. P. Porter, J. Rivkin, and R. M. Kanter, "Competitiveness at the Crossroads: Findings of Harvard Business School's 2012 Survey on U.S. Competitiveness," Harvard Business School, February 2013, www.hbs.edu/competitiveness/pdf/competitiveness-at-a-crossroads.pdf.

25. S. Fleck, J. Glaser, and S. Sprague, "The Compensation-Productivity Gap: A Visual Essay," *Monthly Labor Review*, January 2011, https://www.bls.gov/opub/mlr/2011/01/art3full.pdf; S. Denning, "Debunking Myths About Worker Passion," *Forbes.com*, October 8, 2014, https://www.forbes.com/sites/stevedenning/2014/10/08/debunking-myths-about-worker-passion/.

26. J. Plender, "Blowing the Whistle on Buybacks and Value Destruction," *Financial Times*, March 1, 2016, https://www.ft.com/content/0b71ca32-df0b-11e5-b67f-a61732c1d025.

27. O. Lobel, "Companies Compete but Won't Let Their Workers Do the Same," *New York Times*, May 4, 2017, https://www.nytimes.com/2017/05/04/opinion/noncompete-agreements-workers.html.

28. Denning, "Debunking Myths About Worker Passion."

29. J. Clifton, "Workplace Disruption: From Annual Reviews to Coaching," *Gallup.com*, February 15, 2017, http://www.gallup.com/opinion/chairman/203876/workplace-disruption-annual-reviews-coaching.aspx.

30. S. Denning, "Will We Ever Trust Bankers Again?," *Forbes.com*, February 6, 2013, https://www.forbes.com/sites/stevedenning/2013/02/06/will-we-ever-trust-bankers-again/. The derivatives market is often estimated at more than $1 quadrillion, or more than ten times the size of the total world gross domestic product; see J. B. Maverick, "How Big Is the Derivatives Market?," *Investopedia*, May 27, 2015, http://www.investopedia.com/ask/answers/052715/how-big-derivatives-market.asp.

31. "Banks Have Paid $321 Billion in Fines Since the Financial Crisis," Reuters, March 2, 2017, http://fortune.com/2017/03/03/bank-fines-2008-financial-crisis/.

32. "Five Banks to Plead Guilty to Global Currency Manipulation," *NBC News*, May 20 2015, http://www.nbcnews.com/business/markets/five-banks-plead-guilty-global-currency-manipulation-n361921.

33. A. Viswanatha, "J. P. Morgan to Pay $307 Million over Client 'Steering,'" December 18, 2015, http://www.wsj.com/articles/j-p-morgan-to-pay-367-million-over-client-steering-1450457616; S. Denning,

"Can the Big Banks Get on a Better Track?," *Forbes.com*, December 26, 2015, https://www.forbes.com/sites/stevedenning/2015/12/26/ can-the-big-banks-get-on-a-better-track/.

34. Henning, "When Money Gets in the Way of Corporate Ethics"; K. Mehrotra, L. J. Keller, and E. Pettersson, "Wells Fargo Reaches $110 Million Fake Accounts Settlement," Bloomberg, March 28, 2017, https://www.bloomberg.com/news/articles/2017-03-28/wells-fargo- reaches-110-million-settlement-over-fake-accounts.

35. S. Polk, "What's Wrong with Wall Street," *PBS NewsHour*, August 23, 2016, http://www.pbs.org/newshour/making-sense/whats-wrong-wall- street-culture-breeds-greed/. While being careful not to demonize all bankers, we should also be careful not to dismiss the horrific scale of the missteps. Thus Nobel Prize–winning economist Robert Shiller writes in his book *Finance and the Good Society* about the 2008 crisis: "It is hard to blame the crisis on a sudden outbreak of malevolence. The situation during the boom that created the crisis was rather more like that on a highway where most cars are going just a little too much over the speed limit. In that situation, well-meaning drivers will just flow with the traffic. The U.S. Financial Crisis Inquiry Commission, in its final 2011 report, described the boom as 'madness,' but, whatever it was, it was not for the most part criminal." Yet with banks being convicted of felonies with penalties amounting to more than $300 billion, the missteps go significantly beyond "cars going a little over the speed limit."

36. N. D. Schwartz, "How Wall Street Bent Steel," *New York Times*, December 6, 2014, http://www.nytimes.com/2014/12/07/business/ timken-bows-to-investors-and-splits-in-two.html.

37. Relational ceased doing business in December 2015. See D. Benoit, "Relational Investors Closes Out Portfolio: Move Marks End of Pioneer Activist-Investing Firm," *Wall Street Journal*, February 15, 2016, https://www.wsj.com/articles/relational-investors-closes-out-portfolio- 1455315877.

38. Benoit, "Relational Investors Closes Out Portfolio."

39. Ibid.

40. Timken (TKR) share price as of April 2017 was $43; as of mid-2014: $68. That is a fall of 39 percent. TimkenSteel (TMST) share price as of April 2017 was $14; as of mid-2014: $50. That is a fall of 72 percent.

41. Schwartz, "How Wall Street Bent Steel."

42. Ibid.

43. Sullivan & Cromwell LLP, "2016 U.S. Shareholder Activism Review and Analysis," November 28, 2016, 11, https://sullcrom.com/siteFiles/ Publications/SC_Publication_2016_U.S._Shareholder_Activism_ Review_and_Analysis.pdf.

44. Ibid.

45. Ibid.

46. A. Davis and L. Mishel, "CEO Pay Continues to Rise as Typical Workers Are Paid Less," Economic Policy Institute, June 12, 2014, http://www.epi.org/publication/ceo-pay-continues-to-rise/.

47. W. Lazonick and M. Hopkins, "Corporate Executives Are Making Way More Money Than Anybody Reports," *The Atlantic*, September 15, 2016, https://www.theatlantic.com/business/archive/2016/09/executives-making-way-more-than-reported/499850/; W. Lazonick and M. Hopkins, "If the S.E.C. Measured CEO Pay Packages Properly, They Would Look Even More Outrageous," *Harvard Business Review*, December 22, 2016, https://hbr.org/2016/12/if-the-sec-measured-ceo-pay-packages-properly-they-would-look-even-more-outrageous.

48. J. M. Rose, A. M. Rose, C. Norman, and C. R. Mazza, "Will Disclosure of Friendship Ties Between Directors and CEOs Yield Perverse Effects?," *Accounting Review* 89, no. 4 (July 2014): 1545–1563, http://aaajournals.org/doi/abs/10.2308/accr-50734.

49. Denning, "How the World's Dumbest Idea Killed the Economic Recovery."

50. R. Harding, "Corporate Investment: A Mysterious Divergence," *Financial Times*, July 24, 2013, http://www.ft.com/intl/cms/s/0/8177af34-eb21-11e2-bfdb-00144feabdc0.html.

51. J. Hagel, J. S. Brown, M. Wooll, and A. de Maar, "The Paradox of Flows: Can Hope Flow from Fear?," Deloitte University Press, December 13, 2016, https://dupress.deloitte.com/dup-us-en/topics/strategy/shift-index.html; S. Denning, "The Shift Index 2016: Why Can't U.S. Firms Innovate?," *Forbes.com*, December 15, 2016, https://www.forbes.com/sites/stevedenning/2016/12/15/shift-index-2016-shows-continuing-decline-in-performance-of-us-firms/.

52. N. Smith, "How Finance Came to Dominate the Economy," Bloomberg, April 20, 2016, https://www.bloomberg.com/view/articles/2016-04-20/how-finance-came-to-dominate-the-u-s-economy.

53. How big is too big? In 2012, an International Monetary Fund study showed that "once the [financial] sector becomes too large—when private-sector credit reaches 80 percent to 100 percent of GDP—it actually inhibits growth and increases volatility. In the United States in 2012, private-sector credit was 184 percent of GDP." See J. Arcand, E. Berkes, and U. Panizza, "Too Much Finance," IMF Working Paper WP/12/161, June 2012, https://www.imf.org/external/pubs/ft/wp/2012/wp12161.pdf.

54. Arcand, Berkes, and Panizza, "Too Much Finance." It is sometimes questioned whether the economy is really declining. We are talking here about broad trends over more than half a century, during which

the financial sector has grown significantly. "At its peak in 2006, the
financial services sector contributed 8.3 percent to U.S. GDP, compared to 4.9 percent in 1980 and 2.8 percent in 1950"; see T. Taylor,
"Why Did the U.S. Financial Sector Grow?," *Conversable Economist*
(blog), May 15, 2013, http://conversableeconomist.blogspot.com/
013/05/why-did-us-financial-sector-grow.html. By 2015, the financial
sector was more or less back to what it was in 2006; see N. Irwin,
"Wall Street Is Back, Almost as Big as Ever," *New York Times*, May
18, 2015, http://www.nytimes.com/2015/05/19/upshot/wall-street-
is-back-almost-as-big-as-ever.html. "Between 1947 and 1974, GDP
rose by about four per cent a year, on average, and many American
households enjoyed a surge in living standards. In the nineteen-eighties
and nineties, growth dropped a bit, but still averaged more than three
per cent. Since 2001, however, the rate of expansion has fallen below
two per cent—less than half the postwar rate—and many economists
believe that it will stay there, or fall even further. In economic-policy
circles, the phrase of the moment is 'secular stagnation.'" See
J. Cassidy, "Printing Money," *The New Yorker*, November 23, 2015,
http://www.newyorker.com/magazine/2015/11/23/printing-money-
books-john-cassidy. Accordingly, in terms of current macro statistics,
the growth of the financial sector and the decline of the economy
overall are confirmed. An argument is sometimes made that statistics like GDP don't capture the gains being made by consumers. A
television set that used to cost thousands of dollars now costs hundreds of dollars. True. But in other areas, such as finance, education,
and health, the opposite has occurred: Expenditures have increased
without any obvious gain in outcomes. Much of the financial sector's
growth consists of zero-sum gambling games, with no benefit to the
real economy, yet this shows up as a benefit in the GDP. A law degree
costs four times what it cost three decades ago, but is not noticeably
better, yet this shows as a gain in GDP. Massive high-tech expenditures
in medicine (which count as benefits) often yield only modest, zero, or
even negative gains in terms of quality or length of life. The net of all
these pluses and minuses has yet to be calculated. At this point, there
is no basis for assuming a priori that the overcalculations exceed the
undercalculations.

55. Bower and Paine, "The Error at the Heart of Corporate Leadership."
56. "Analyse This: The Enduring Power of the Biggest Idea in Business,"
    The Economist, March 31, 2016, http://www.economist.com/news/
    business/21695940-enduring-power-biggest-idea-business-analyse.
57. "Analyse This," *The Economist*; See also S. Denning, "The Econo-
    mist Defends the World's Dumbest Idea," *Forbes.com*, April 3, 2016,

https://www.forbes.com/sites/stevedenning/2016/04/03/the-economist-defends-the-worlds-dumbest-idea/.

58. S. Denning, "The Creative Economy in France, *Forbes.com*, July 13, 2014, http://www.forbes.com/sites/stevedenning/2014/07/13/the-creative-economy-in-france-givenchy-vinci/.

59. "Jack Ma Brings Alibaba to the U.S.," interview by Lara Logan, *CBS*, September 28, 2014, http://www.cbsnews.com/news/alibaba-chairman-jack-ma-brings-company-to-america/.

60. E. Reguly, "Maybe It's Time for CEOs to Put Shareholders Second," *Globe and Daily Mail*, last updated September 27, 2013, http://www.theglobeandmail.com/report-on-business/rob-magazine/maybe-its-time-for-ceos-to-put-shareholders-second/article14507016/.

61. S. Denning, "The New Management Paradigm: John Mackey's Whole Foods," *Forbes.com*, January 5, 2013, http://www.forbes.com/sites/stevedenning/2013/01/05/the-new-management-paradigm-john-mackeys-whole-foods/. As of mid-2017, Whole Foods was under pressure from the hedge fund Jana and the money manager Neuberger Berman to accelerate change; see L. Thomas, "Buying into the Turmoil: Investors Embrace the Risks," *New York Times*, May 10, 2017, https://www.nytimes.com/2017/05/10/business/dealbook/whole-foods-board.html.

62. M. Benioff, "A Call for Stakeholder Activists," *Huffington Post*, February 2, 2015, http://www.huffingtonpost.com/marc-benioff/a-call-for-stakeholder-activists_b_6599000.html.

63. "BlackRock CEO Larry Fink tells the world's biggest business leaders to stop worrying about short-term results," *BusinessInsider.com*, April 14, 2015, http://www.businessinsider.com/larry-fink-letter-to-ceos-2015-4.

64. S. Denning, "How CEOs Became Takers, Not Makers"; S. Denning, "From CEO Takers to CEO Makers," *Forbes.com*, August 20, 2014, http://www.forbes.com/sites/stevedenning/2014/08/20/from-ceo-takers-to-ceo-makers-the-great-transformation/; S. Denning, "The Economist: Blue Chips Are Addicted to Corporate Cocaine," *Forbes.com*, September 19, 2014, http:/www.forbes.com/sites/stevedenning/2014/09/9/the-economist-blue-chips-are-addicted-to-corporate-cocaine/.

CHAPTER 9

1. "Corporate cocaine: Companies are spending record amounts on buying back their own shares. Investors should be worried," *The Economist*, September 13, 2014, http://www.economist.com/news/leaders/21616950-companies-are-spending-record-amounts-buying-back-their-own-shares-investors-should-be; "The repurchase revolution: Companies have been gobbling up their own shares at an exceptional rate: there are good

reasons," *The Economist*, September 12, 2014, http://www.economist.com/news/business/21616968-companies-have-been-gobbling-up-their-own-shares-exceptional-rate-there-are-good-reasons.

2. W. Lazonick, "Profits Without Prosperity," *Harvard Business Review*, September 2014.

3. S. Denning, "The Economist, Blue Chips Are Addicted to Corporate Cocaine," *Forbes.com*, April 19, 2014, http://www.forbes.com/sites/stevedenning/2014/09/19/the-economist-blue-chips-are-addicted-to-corporate-cocaine/; K. Brettell, D. Gaffen, and D. Rohde, "The Canni-balized Company: How the Cult of Shareholder Value Has Reshaped Corporate America; A Special Report," Reuters, November 16, 2015, http://www.reuters.com/investigates/special-report/usa-buybacks-cannibalized; J. Plender, "Blowing the Whistle on Buybacks and Value Destruction," *Financial Times*, March 1, 2016, https://www.ft.com/content/0b71ca32-df0b-11e5-b67f-a61732c1d025; S. Denning, "The Best Management Article of 2014," *Forbes.com*, March 26, 2015, https://www.forbes.com/sites/stevedenning/2015/03/26/the-best-management-article-of-2014/.

4. W. Lazonick, "How Stock Buybacks Make Americans Vulnerable to Globalization," Working Paper, East-West Center Workshop on Mega-Regionalism: New Challenges for Trade and Innovation, March 11, 2016, https://papers.ssrn.com/sol3/papers.cfm?abstract_id=2745387.

5. Ibid.

6. Ibid.

7. Ibid.

8. Plender, "Blowing the Whistle on Buybacks and Value Destruction." By the first quarter of 2017, the stock market had soared without foun-dation to such extravagant highs that even corporate boards began to show some moderation in the rush to extract value; see "Market Too Frothy for Buybacks: CFOs and Boards," *Seeking Alpha*, May 2, 2017, https://seekingalpha.com/article/4067753-market-frothy-buybacks-cfos-boards.

9. Lazonick, "How Stock Buybacks Make Americans Vulnerable to Globalization."

10. Jeff Cox, "This 'investor obsession' during the bull market is falling out of fashion," *CNBC*, May 2, 2017, http://www.cnbc.com/2017/05/01/buyback-obsession-during-the-bull-market-is-falling-out-of-fashion.html.

11. J. Bezos: "Letter to Amazon Shareholders," 1997, http://media.corporate-ir.net/media_files/irol/97/97664/reports/Shareholderletter97.pdf; S. Denning, "How Not to Reclaim the World's Dumbest Idea,"

Forbes.com, August 31, 2016, https://www.forbes.com/sites/stevedenning/2016/08/31/hbr-how-not-to-reclaim-the-worlds-dumbest-idea/.

12. Bezos, "Letter to Amazon Shareholders."

13. A. Boynton, "Unilever's Paul Polman: CEOs Can't Be Slaves to Shareholders," July 20, 2015, https://www.forbes.com/sites/andyboynton/2015/07/20/unilevers-paul-polman-ceos-cant-be-slaves-to-shareholders/.

14. S. Denning, "Should Wall Street Reward Adobe's Failing Profits," *Forbes.com*, March 28, 2014, http://www.forbes.com/sites/stevedenning/2014/03/28/should-wall-street-reward-adobes-falling-profits/.

15. S. Denning, "Shift Index 2016: Why Can't Firms Innovate?," *Forbes.com*, December 15, 2016, https://www.forbes.com/sites/stevedenning/2016/12/15/shift-index-2016-shows-continuing-decline-in-performance-of-us-firms/.

CHAPTER 10

1. J. McManus, "The Risk of Too Much Cost Focus: Saving on expense may be an upside of tech innovation and investment, but the real goal should be creating value, not merely increasing efficiency," *BuilderOnLine*, April 2, 2017, http://www.builderonline.com/builder-100/strategy/the-risk-of-too-much-cost-focus_o.

2. R. Coase, "The Nature of the Firm," *Economica* 4, no. 16 (November 1937): 386–405, http://onlinelibrary.wiley.com/doi/10.1111/j.1468-03351937.tb00002.x/abstract.

3. A. Cockcroft, "Evolution of Business Logic from Monoliths Through Microservices to Functions," February 16, 2017, https://read.acloud.guru/evolution-of-business-logic-from-monoliths-through-microservices-to-functions-ff464b95a44d. "In the past, costs were high and efficiency concerns dominated, with high time to value regarded as the normal state of affairs. . . . When costs dominate, that's where the focus is, but as costs reduce and software impact increases, the focus flips towards getting the return earlier."

4. C. M. Christensen, J. H. Grossman, and J. Hwang, *The Innovator's Prescription: A Disruptive Solution for Health Care* (New York: McGraw-Hill Education, 2008).

5. Ibid., 263.

6. Ibid., 265.

7. S. Denning, "Why Amazon Can't Make a Kindle in the USA," *Forbes.com*, August 17, 2011, http://www.forbes.com/sites/stevedenning/2011/08/17/why-amazon-cant-make-a-kindle-in-the-usa/.

8. P. F. Drucker, *The Practice of Management* (New York: HarperCollins, 1954), 65. "Company after company is working on the definition

of the key areas, on thinking through what should be measured and on fashioning the tools of measurement. Within a few years our knowledge of what to measure and our ability to do so should therefore be greatly increased." See also P. Drucker and R. Wartzman, *The Drucker Lectures: Essential Lessons on Management, Society and Economy* (New York: McGraw-Hill Education, 2010), 62. Overall, being "data-informed" may be better approach than being "data-driven"; see the earlier discussion in Chapter 5.

9. "Total Cost of Ownership Estimator," *Reshoring Initiative*, http://www .reshorenow.org/TCO_Estimator.cfm.

10. H. Moser conversation with S. Denning, "What Went Wrong at Boeing?," *Forbes.com*, January 17, 2013, https://www.forbes.com/sites/ stevedenning/2013/01/17/the-boeing-debacle-seven-lessons-every-ceo-must-learn/. See also "Offshoring: What's the Total Cost of Ownership?," *Quality Digest*, September 21, 2011, https://www.qualitydigest. com/inside/quality-insider-article/offshoring-what-s-total-cost-ownership.html.

11. Denning, "What Went Wrong at Boeing."

12. G. P. Pisano and W. C. Shih, "Restoring American Competitiveness," *Harvard Business Review*, July–August 2009, https://hbr.org/2009/07/ restoring-american-competitiveness.

13. Ibid. The components of a Kindle are manufactured in Asia. The flex circuit connectors, the highly polished injection-molded case, the controller board, and the lithium polymer battery are made in China. The electrophoretic display is made in Taiwan. The wireless card is made in South Korea.

14. The list of sectors already lost includes: "fabless chips; compact fluorescent lighting; LCDs for monitors, TVs, and handheld devices like mobile phones; electrophoretic displays; lithium ion, lithium polymer, and NiMH batteries; advanced rechargeable batteries for hybrid vehicles; crystalline and polycrystalline silicon solar cells, inverters, and power semiconductors for solar panels; desktop, notebook, and netbook PCs; low-end servers; hard-disk drives; consumer networking gear such as routers, access points, and home set-top boxes; advanced composite [materials] used in sporting goods and other consumer gear; advanced ceramics and integrated circuit packaging."

15. C. Fishman, "The Insourcing Boom," *The Atlantic*, December 2012, http://www.theatlantic.com/magazine/archive/2012/12/the-insourcing-boom/309166/.

16. Ibid. See also S. Denning, "Why Apple and GE Are Bringing Back Manufacturing," *Forbes.com*, December 7, 2012, https://www.forbes .com/sites/stevedenning/2012/12/07/why-apple-and-ge-are-bringing-manufacturing-back/.

17. In 2012, the China-made GeoSpring retailed for $1,599. The Louis-ville-made GeoSpring retails for $1,299 (Fishman, "The Insourcing Boom"). By 2015, GE had raised the price to $1,900, according to pricing data at the GE Appliances website (http://products.geappliances. com/appliance/dealer-locations/GEH80DFEJSR; http://www .geappliances.com/ge/heat-pump-hot-water-heater.htm).

18. Fishman, "The Insourcing Boom."

19. Ibid.

20. In June 2016, GE announced that it was selling its appliance division to Haier, the world's leading appliance manufacturer, which is based in Qingdao, China. A. C. Thompson, "It's Official: GE Appliances Belongs to Haier," *CNET*, June 6, 2016, https://www.cnet.com/news/ its-official-ge-appliances-belongs-to-haier/.

21. J. Freeman et al., "Theory of Constraints and Throughput Accounting," Chartered Institute of Management Accountants (CIMA) *Topic Gateway Series No. 26*, March 2007, http://www.cimaglobal. com/Documents/ImportedDocuments/26_Theory_of_Constraints_ and_Throughput_Accounting.pdf.

CHAPTER 11

1. J. Dewey, *The Collected Works of John Dewey* v. 14; 1939-1941 (Amsterdam Netherlands: Pergamon Media/Elsevier, 2015), 107.

2. P. Gorski, "Michael Porter Is Bankrupt AND the Framework of a Blindfolded Chimpanzee," *Gorski Ventures News*, November 13, 2012, http://www.gorskiventures.com/michael-porter-is-bankrupt-and-the-framework-of-blindfolded-chimpanzee/.

3. Ibid. Peter Gorski suggested that "even a blindfolded chimpanzee throwing darts at the Porter Five Forces framework can select a busi-ness strategy that performs as well as that prescribed by Dr. Porter and other high-paid strategy consultants." See also S. Denning, "What Killed Michael Porter's Monitor Group? The One Force That Really Matters," *Forbes.com*, November 20, 2012, https://www.forbes.com/ sites/stevedenning/2012/11/20/what-killed-michael-porters-monitor-group-the-one-force-that-really-matters/; S. Denning, "Even Monitor Didn't Believe in the Five Forces," *Forbes.com*, November 24, 2012, https://www.forbes.com/sites/stevedenning/2012/11/24/even-monitor-didnt-believe-in-the-five-forces/; S. Denning, "It's Official: The End of Competitive Advantage," *Forbes.com*, June 2, 2013, https://www .forbes.com/sites/stevedenning/2013/06/02/its-official-the-end-of-competitive-advantage/. See also the counter-critique from the strategy faculty at the Kellogg School of Management at Northwestern Uni-versity: J. Love, "The End of Strategy? Our Faculty Discusses,"

January 7, 2013, https://insight.kellogg.northwestern.edu/blogs/entry/the_end_of_strategy_our_faculty_discusses.

4. M. E. Porter, *Competitive Strategy* (New York: Free Press, 1980). The book was voted the ninth most influential management book of the twentieth century in a poll of the Fellows of the Academy of Management.

5. M. Stewart, *The Management Myth: Why the Experts Keep Getting it Wrong* (New York: W. W. Norton & Company, 2009), 160. Stewart is a consulting insider, and his book is enlightening, but misleadingly titled.

6. Ibid., 168.

7. J. Magretta, *Understanding Michael Porter* (Boston: Harvard Business School Publishing, 2012), 8.

8. Stewart, *The Management Myth*, 207.

9. C. von Clausewitz, *On War*, published posthumously in 1832; L. Freedman, *Strategy: A History* (Oxford, U.K.: Oxford University Press, 2013), 87.

10. H. G. von Moltke, *Militarische Werke*, vol. 2, part 2. Reproduced in D. J. Hughes, ed., *Moltke on the Art of War: selected writings* (New York: Presidio Press, 1993), 45.

11. *Mission Command: Command and Control of Army Forces*, Field Manual No. 6-0, August 2003.

12. Conceptually, detailed command and mission command reflect two different approaches to dealing with uncertainty. *Detailed command* is data-based and information-focused. It aims to reduce uncertainty at the higher echelons by collecting more and better data and increasing the information processing capability. It trades off speed of action for completeness of information. *Mission command,* by contrast, is action-oriented. It aims at reducing uncertainty evenly throughout the organization. Leaders educate their organizations to codevelop a widely understood strategic vision and manage a set of strategic initiatives as part of normal operations. They delegate authority for decision making to those levels that can acquire and process information and move into action quickly without waiting for detailed orders. The process makes full use of the organization's talent. See *Mission Command: Command and Control of Army Forces*, para. 1–50.

13. A. Murray, "The End of Management," *Wall Street Journal*, August 21, 2010, http://www.wsj.com/articles/SB10001424052748704476104575439723695579664.

14. M. Porter, "How Competitive Forces Shape Strategy," *Harvard Business Review*, March 1979, https://hbr.org/1979/03/how-competitive-forces-shape-strategy/ar/1, and republished as "The Five Competitive

Forces That Shape Strategy," *Harvard Business Review*, January 2008, https://hbr.org/2008/01/the-five-competitive-forces-that-shape-strategy.

15. In the 2008 version, the sentence has only been slightly amended to read: "In essence, the job of the strategist is to understand and cope with competition."

16. Porter, "How Competitive Forces Shape Strategy." In the 2008 republished version of the original article, the language is somewhat modified: "Understanding the competitive forces, and their underlying causes, reveals the roots of an industry's current profitability while providing a framework for anticipating and influencing competition and profitability over time." Thus "the ultimate profit potential" has been modified to become "the roots of an industry's current profitability," and "the framework for anticipating influencing . . . profitability over time." The problem here is that the thinking is still the same. An industry's current profitability is a very poor framework for anticipating future profitability, given massive disruptive innovation now under way across all sectors. In any event, there is no acknowledgement of the change in thinking. The republished article is presented as embodying the "timeless truths" of Porter's thinking.

17. Magretta, *Understanding Michael Porter*, 31.

18. Stewart, *The Management Myth,* 183.

19. Ibid., 182.

20. Ibid., 183.

21. J.B. Quinn, *Strategies for change: Logical incrementalism* (Homewood, IL: R.D. Irwin, 1980), 122.

22. Denning, "Even Monitor Didn't Believe in the Five Forces"; see also the Kellogg School of Management's counter-critique by Love, "The End of Strategy? Our Faculty Discusses."

23. The idea that strategy is innovation was discussed in Gary Hamel, "Strategy as Revolution," *Harvard Business Review*, July–August 1996, https://hbr.org/1996/07/strategy-as-revolution.

CHAPTER 12

1. T. Herzl, *Old New Land* (Leipzig: Seemann Nachf, 1902), 296: "But if you do not wish it, all this that I have related to you is and will remain a dream." The simplified, more positive version of the phrase, "If you will it, it is no dream," became a popular slogan of the Zionist movement—the striving for a Jewish National Home in Israel.

2. S. Denning, "Coding: Agile and Scrum Go Mainstream," *Forbes.com*, June 14, 2015, https://www.forbes.com/sites/stevedenning/2015/06/14/coding-agile-scrum-go-mainstream/.

3. D. K. Rigby, J. Sutherland, and H. Takeuchi, "Embracing Agile," *Harvard Business Review*, April 2016; S. Denning, "HBR's Embrace

of Agile," *Forbes.com*, April 21, 2016, http://www.forbes.com/sites/stevedenning/2016/04/21/hbrs-embrace-of-agile/; K. Brettell, D. Gaffen, and D. Rohde, "The Cannibalized Company," Reuters, November 16, 2015, http://www.reuters.com/investigates/special-report/usa-buybacks-cannibalized/; S. Denning, "How Corporate America Is Cannibalizing Itself," *Forbes.com*, November 18, 2015, https://www.forbes.com/sites/stevedenning/2015/11/18/how-corporate-america-is-cannibalizing-itself/; S. Denning, "Financial Times Slams Share Buybacks," *Forbes.com,* March 22, 2016, https://www.forbes.com/sites/stevedenning/2016/03/22/financial-times-slams-share-buybacks/; S. Denning, "Resisting the Lure of Short-Termism," *Forbes.com*, January 8, 2017, https://www.forbes.com/sites/stevedenning/2017/01/08/resisting-the-lure-of-short-termism-how-to-achieve-long-term-growth/.

4. J. Cox, "This 'Investor Obsession' During the Bull Market Is Falling Out of Fashion," *CNBC*, May 1, 2017, http://www.cnbc.com/2017 05/01/buyback-obsession-during-the-bull-market-is-falling-out-of-fashion.html.

5. Chapter 12 builds on Carlota Perez's brilliant book, *Technological Revolutions and Financial Capital: The Dynamics of Bubbles and Golden Ages* (Cheltenham, U.K.: Edward Elgar, 2003). As W. Brian Arthur of the Santa Fe Institute in New Mexico wrote: "Before I read this book I thought that the history of technology was—to borrow Churchill's phrase—merely 'one damned thing after another.' Not so. Carlota Perez shows us that historically technological revolutions arrive with remarkable regularity, and that economies react to them in predictable phases." See also S. Denning, "Understanding Disruption: Insights from the History of Business," *Forbes.com*, June 24, 2014, https://www.forbes.com/sites/stevedenning/2014/06/24/understanding-disruption-insights-from-the-history-of-business/.

6. This quote is often attributed to Mark Twain, although there is no documented proof of that. For a further investigation of the provenance of this adage, see "History Does Not Repeat Itself, But It Rhymes," *Quote Investigator*, http://quoteinvestigator.com/2014/01/12/history-rhymes/.

7. See for example Nazi Germany, Russia, and the Soviet Union.

8. C. Perez, "The double bubble at the turn of the Century: Technological roots and structural implications," Centre for Financial Analysis & Policy, Judge Business School, Cambridge University, U.K., CFAP/CERF Working Paper No. 31: 22.

9. F. Partnoy and J. Eisinger, "What's Inside America's Banks?," *The Atlantic*, January/February 2013, https://www.theatlantic.com/magazine/archive/2013/01/whats-inside-americas-banks/309196/.

10. S. Denning, "Lest We Forget: Why We Had a Financial Crisis," *Forbes.com*, November 22, 2011, https://www.forbes.com/sites/stevedenning/2011/11/22/5086/.

11. The defeat of the powerful U.S. chemical companies who opposed the regulation of pesticides in the 1960s is an interesting illustration of how a combination of knowledge and courage can overcome money. See Rachel Carson, *Silent Spring* (New York: Houghton Mifflin Harcourt, 1962).

12. These are examples of "big ideas" presented at a recent international TED conference. See C. Itkowitz, "Prioritizing These Three Things Will Improve Your Life—and Maybe Even Save It," *Washington Post*, April 28, 2017, https://www.washingtonpost.com/news/inspired-life/wp/2017/04/28/prioritizing-these-three-things-will-improve-your-life-and-maybe-even-save-it/.

13. A. Cockcroft, "Evolution of Business Logic from Monoliths Through Microservices to Functions," *A Cloud Guru*, February 16, 2017, https://read.acloud.guru/evolution-of-business-logic-from-monoliths-through-microservices-to-functions-ff464b95a44d.

14. Ibid.

15. Quoted in Daniel Altman, "Managing Globalization: Q & A with Joseph Stiglitz," *International Herald Tribune,* October 11, 2006. See also J. Schlefer, "There Is No Invisible Hand," *Harvard Business Review*, April 2004, https://hbr.org/2012/04/there-is-no-invisible-hand; T. Worstall, "The Death of Macroeconomics: There Is No Invisible Hand," *Forbes.com*, April 2012, https://www.forbes.com/sites/imworstall/2012/04/12/the-death-of-macroeconomics-there-is-no-invisible-hand/.

16. Rigby, Sutherland, and Takeuchi, "Embracing Agile"; Denning, "HBR's Embrace of Agile."

17. D. Brooks, "This Age of Wonkery," *New York Times*, April 11, 2017, https://www.nytimes.com/2017/04/11/opinion/this-age-of-wonkery.html.

18. Under the courageous thought leadership of its editor, Robert Randall, *Strategy & Leadership* journal (http://www.emeraldinsight.com/loi/sl -- subscription required) has published more than thirty articles on Agile management since 2010.

19. "The true elite of modern societies is composed of engineers, mechanics, and artisans—masters of reality, not big thinkers." A. Gopnik, "Are Liberals on the Wrong Side of History?," *The New Yorker*, March 20, 2017, http://www.newyorker.com/magazine/2017/03/20/are-liberals-on-the-wrong-side-of-history.

20. The first few decades of the fifteenth century is seen by some critics as a moment in history when civilization was at a peak, a Renaissance of the human spirit. See K. Clark, *Civilization* (New York: Harper & Row, 1969). Like today, there was much to criticize. Crass materialism. A love of making money. Concentration of wealth in the hands of a few. Financial meltdowns. Political chicanery. Environmental disasters. Raucous, uncouth marketplaces occupied by self-seeking bankers, merchants, and traders. Yet there was also much to admire. Among the greatest achievements was that things were kept on a human scale. Some are disappointed when they visit the famous beginnings of Renaissance architecture, like the Pazzi Chapel in Florence or the Old Sacristy of San Lorenzo, because they seem so small. "They don't try to impress us or crush us by size and weight, as all God-directed architecture does. Everything is adjusted to the scale of reasonable human necessity. They are intended to make each individual more conscious of his powers, as a complete moral and intellectual being. They are an assertion of the dignity of man" (*Civilization*, 64). They have the qualities of a mathematical theorem: "clarity, economy, elegance."

INDEX

Note: Page numbers in italics indicate figures.

ABB, 94
accountability, 87
accounting, 214–15, *215*
action orientation, 92, 130
"activist hedge funds," 169, 174–76, 184
activist shareholders, 174–76, 198
adaptability, xiii–xiv, xvii
agency problem, 170, 176–77
agenda for action, 255–64
"Age of Agile," 275n15
Agile
  definition of, 23
  at scale, 103–17
  understanding, 13–14
Agile leaders
  share buybacks and, 199–200
  stock market and, 199–200
Agile Loans, 9
Agile management, 21, 201–2, 205–6, 246
  accounting and, 214–15, *215*
  before Agile Manifesto, 274n7
  at Cerner Corporation, 73–80
  commitment of, 98
  competitive advantage and, 223–24, 231–32
  economic critique offered by, 247
  explaining to a CFO, 213
  financial critique offered by, 246
  goal of, 255
  hierarchy and, 278–79n20
  implementation of, 33, 97–101, 103–17
  implications of, 61
  legal critique offered by, 246
  in manufacturing, 274–75n9
  misunderstandings of, 13
  moral critique offered by, 247
  "New Age" talk and, 278–79n20
  operating under three laws, 21–22
  people management and, 73–80
  philosophical critique offered by, 247
  political critique offered by, 247
  practical critique offered by, 246
  practices, 33–37
  protecting from shareholder value thinking, 179
  at scale, 103–17
  spread of, 10–11
  strategy and, 224
  teams and, 46–47, *46*
  vs. traditional management, 240–41
Agile managers
  within the corporation, 201–2
  mastering role of, 108
Agile Manifesto, xvi, 22–24, 33, 205, 248, 274n8
  Agile management before, 274n7
  first principle of, 16
Agile movement, 14–16, 18–20, 94
Agile Onboarding, 9
Agile organizations, 19, 89
  vs. bureaucratic organizations, 17–18, *17*
  hierarchy in, 19–20
Agile software development, manifesto for, 22–23
Agile teams, 89–90, 94. *See also* teams
agility, market-based approaches, 94–95
Airbnb, xiv, 97, 125
Alcoholics Anonymous, 91, 93
alignment, autonomy and, 106–7
Amazon, 57, 129, 134, 200, 201, 229, 254, 257, 273n2
  Kindle, 211, 297n13
Anthony, Scott, 119–20
anthropologists, 42
"anti-lean-startups" movement, 285n9
Apache Software Foundation, 80
Apple, xiii, 29, 45, 122, 129, 130, 134, 229, 254, 257
  iPad, 72
  iPhone, 29, 72, 97, 121, 149
  iPod, 124
  Music, 45
  "The Newton," 277n3
  organizational change at, 153–54
approach, 127–33, 145–46, 148–49, 152–53, 155–56
Archstone, 210

Printed in the USA
CPSIA information can be obtained
at www.ICGtesting.com
JSHW030808070624
64397JS00011B/226

9 781400 242405